Special Drilling Operations

by David J. Morris

Published by
Petroleum Extension Service
Division of Continuing Education
The University of Texas at Austin
Austin, Texas
1984

To obtain additional training materials, contact:

PETROLEUM EXTENSION SERVICE
The University of Texas at Austin
J. J. Pickle Research Campus, R8100
Austin, TX 78712

Telephone: (512) 471-5940
or 1-800-687-4132
FAX: (512) 471-9410
or 1-800-687-7839

© 1984 by The University of Texas at Austin
All rights reserved
Published 1984. Fifth Impression 2003
Printed in the United States of America

This book or parts thereof may not be reproduced in any form without permission of Petroleum Extension Service, The University of Texas at Austin.

Brand names, company names, trademarks, or other identifying symbols appearing in illustrations or text are used for educational purposes only and do not constitute an endorsement by the author or publisher.

The University of Texas at Austin is an equal opportunity institution. No state tax funds were used to print or mail this publication.

Catalog No. 2.01310
ISBN 0-88698-145-X

CONTENTS

Preface v
Acknowledgments vii

Lesson 1: Controlled Directional Drilling, Part I 3
    Introduction 3
    Directional Wells 5
    Directional Surveying 12
    Questions 29

Lesson 2: Controlled Directional Drilling, Part II 35
    Changing the Course of the Hole 37
    Special Problems in Directional Drilling 51
    Summary 56
    Questions 57

Lesson 3: Open-Hole Fishing 61
    Introduction 63
    Causes of Fishing Jobs 63
    Fishing Equipment and Techniques 70
    The Economics of Fishing 106
    Questions 109

Lesson 4: Well Control, Part I 113
    Introduction 115
    Well Pressures 117
    Causes of Kicks 133
    Signs of a Kick 142
    Questions 149

Lesson 5: Well Control, Part II 155
    Controlling Kicks 157
    Preventer Drills 184
    Questions 185

Lesson 6: Optimization 191
    Introduction 193
    Bits 194
    Weight on Bit and Rotary Speed 201
    Drilling Fluids 206
    Bit Hydraulics 209
    Formation Properties 211
    Computerized Optimization 212
    Questions 215

PREFACE

Special Drilling Operations is Segment III of the three-part Drilling Technology Series. These textbooks are intended for industry and college petroleum technology students. The new books are designed to respond to a need for more explanation of drilling procedures and of the math used in connection with those procedures. After reading the text, students can fill out the questions and hand in the pages to the classroom teacher for grading. When the lessons are returned, the student can insert them into his or her binder in order to have a personal manual for future study. (All three books are also available from PETEX as correspondence courses for those students wishing to take the course through the University of Texas.)

Segment I, of this Series, *Introduction to Rotary Drilling,* teaches basic background material important to anyone directly involved in the drilling industry. Segment II, *Routine Drilling Operations,* describes standard operating practices in some detail.

Segment III deals with operations that are not conducted on every rig but are commonly used in special drilling situations. It will be of value to persons familiar with some aspects of routine drilling who want to learn more about certain operations under special circumstances: directional drilling, fishing, well control, and optimization. The mathematical calculations in this segment are explained in the context of appropriate rig procedures; a familiarity with basic algebra will help the student understand the material.

We hope this text and the other two segments will be of value to those students eager to have knowledge of drilling theory presented in the context of practical procedural instructions.

Annes McCann
Technical Writing Supervisor

ACKNOWLEDGMENTS

The material in this textbook is based on the PETEX-IADC Rotary Drilling Series and other appropriate training materials used by drilling contractors. However, many individuals provided new information and illustrations to help make Segment III as current as possible; others contributed their time and expertise to ensure its accuracy. Among these are Ron Bitto, Tom Davis, and Steve McGee of Eastman Whipstock; Bill Carothers, Bob Childs, Ira Davis, and Cowboy Griffith of Wilson Downhole Services; Alan McConnell, Louis Pasche III, and Bill Watson of NL Sperry-Sun; Archie Parker of Cochran-Dean; J. J. Robertson of Gearhart Industries; and Jordan Sawdo of Controlled Drilling Consultants. Special thanks also go to PETEX staff consultants Ron Baker and Dick Donnelly for their help in revising and editing the material.

                                            Jeff Morris
                                            Training Specialist

Lesson 1
Controlled Directional Drilling
(Part I)

Introduction

Directional Wells

Directional Surveying

Lesson 1
CONTROLLED DIRECTIONAL DRILLING, PART I

INTRODUCTION

In Huntington Beach, California, 1930, a drilling contractor proposed to assemble a rig on a long oceanside pier and drill straight down to reach an offshore petroleum deposit. This was common practice at the time; however, local officials, for one reason or another, vetoed his well. The driller, undaunted, rigged up on dry land instead and drilled a slanted hole out beneath the seafloor.

The determined driller did not invent directional drilling; wells had been deviated since 1895 for such purposes as sidetracking equipment stuck in the hole. Moreover, holes that had been assumed to be vertical were sometimes found to be crooked. In Oklahoma in the 1920s, large differences in depth had been noted among wells drilled into the same reservoir. Inclinometer surveys had shown many of the wells to be anything but vertical, in some cases bottoming out far from where they were meant to.

But the Huntington Beach well was the first recorded application of *controlled directional drilling:* deviating a wellbore along a planned course to an underground target located a given horizontal distance from the top of the hole.

Unfortunately, due either to the driller's reputation or to his ingenuity in flouting authority, controlled directional drilling was immediately considered a highly suspicious activity. Indeed, despite its increasingly common and legitimate use in Huntington Beach and elsewhere through the 1930s, the term *slant-hole drilling* often meant someone was being cheated. East Texas was plagued by oil rustlers drilling slanted holes beneath fencelines.

However, East Texas was also the area where another important use of controlled directional drilling was pioneered. In 1934 a well near Conroe ran wild, blowing out a crater that swallowed the rig and all surface equipment. A slanted relief well was drilled to bottom out near the blown-out well. Mud was pumped in under pressure; channeling through the high-pressure formation, it controlled the blowout and saved nearby wells.

In the decades since, controlled directional drilling has proved useful in many other ways. It has allowed safer and more efficient production of onshore reservoirs and has made large-scale offshore drilling economically feasible.

Directional drilling has become a specialty. An oil company operator hires a directional drilling service company to provide directional hole planning, sophisticated directional tools, and on-site assistance. The operator also hires a surveying company to measure the well's trajectory, except when the contract services of the directional drilling company include all necessary surveys.

As soon as the well owner approves the plan, the directional drilling supervisor becomes, in effect, a member of the well crew. At the wellsite, his main job is to help the driller keep the actual wellbore as close as possible to its planned course. This job includes—

1. conducting single-shot directional surveys to determine hole inclination and direction at selected depths;
2. calculating and plotting the course of the hole from directional survey data;
3. helping the driller select appropriate deflection tools to direct the course of the hole;
4. helping the driller orient deflection tools to accomplish desired course changes;
5. specifying bottomhole assemblies needed for directional control; and
6. dealing with the special problems of drilling a directional well.

In this lesson you will learn the types and uses of directional holes, how a well is planned, what information and equipment are needed and how they are obtained and used, and some of the problems associated with directional drilling.

DIRECTIONAL WELLS

Directional drilling has a wide variety of applications. Each hole plan is tailored to the situation. Reservoir location, surface accessibility, formation hardness, and equipment availability all play roles in planning well trajectory. However, most directional wells follow one of three basic patterns.

Basic Well Patterns

The Type I well (fig. 1) is deflected near the surface to a specified angle, then continued to total depth at the same angle. It is often used for moderately deep wells where the oil-bearing rock is a single zone and intermediate casing is not required. In a deeper hole requiring a large offset, intermediate casing may be set through and beyond the curved section and uncased hole drilled at the same angle to total depth.

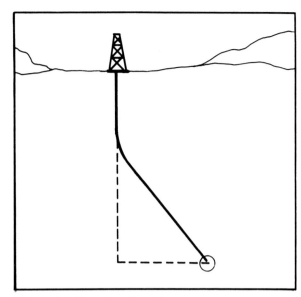

Figure 1.1. Type I well pattern

Type II is an S-shaped hole (fig. 1.2). After the initial deflection, the hole is drilled to a specified drift angle and deflected back to vertical to reach target. Intermediate casing may be set through the second deflection and the rest of the hole drilled vertically to total depth. A Type II hole is used where gas zones, salt water, or other factors call for an intermediate casing string. It is sometimes used to locate a blown-out well to aid in drilling a relief well. It also allows accurate bottomhole spacing where many wells are drilled from the same surface location, as in offshore drilling.

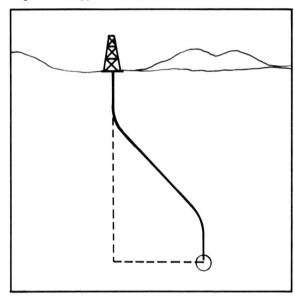

Figure 1.2. Type II well pattern

A Type III well (fig. 1.3) is deflected at greater depth than a Type I or II well, but horizontal deviation is usually less. Hole angle usually continues to build until the target is reached. The deflected part is normally uncased. Type III wells are useful for drilling beneath salt dome overhangs, for multiple exploration holes, and for deep targets.

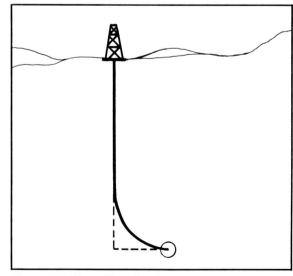

Figure 1.3. Type III well pattern

Applications of Directional Drilling

Directional drilling makes it possible to produce oil from undersea reservoirs far from shore. For efficient production, most petroleum reservoirs require many wells; however, the cost of an offshore production platform is often much greater than the value of the oil or gas that can be produced from a single well. By deviating the wellbores, it is possible to drill many wells from a single platform into different parts of a reservoir, spacing the hole bottoms for optimum recovery (fig. 1.4). Type II wells are particularly useful for accurate spacing.

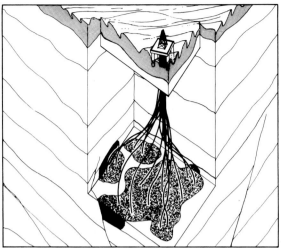

Figure 1.4. Multiple directional wells drilled from an offshore platform

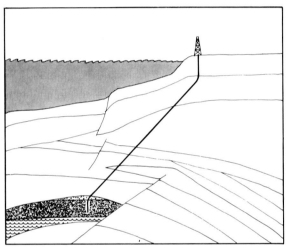

Figure 1.5. Directional hole from onshore rig to offshore reservoir

As in Huntington Beach, many undersea reservoirs are close enough to shore to be reached by land-based deviated wells (fig. 1.5). The well illustrated is a Type II hole, but Type I could also be used here.

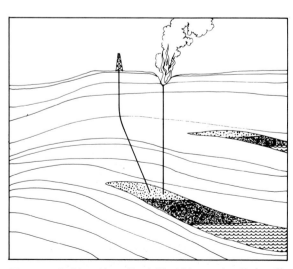

Figure 1.6. Directionally drilled blowout relief well

Since 1934, using directional wells to control blowouts has become commonplace (fig. 1.6). The original relief well worked even though it only bottomed out near the blowout; the technology of the '80s enables a driller to bottom out a relief well within 10 feet of the casing seat of a blown-out well.

Many directional wells are drilled to reach reservoirs inaccessible from a point directly above because of surface obstacles such as hills, lakes, or buildings (fig. 1.7A). Others are drilled to avoid geologic obstructions; the well shown at B was drilled into a salt dome, cemented off, and redirected to the reservoir beneath the overhang. Holes drilled through salt are prone to washout and lost circulation and are often hard to case, so wells are often drilled around salt domes, as shown at C, to avoid both the salt and the hard caprock commonly found atop it. Note that A is a Type I hole, while B and C are both Type III holes.

Drilling through a fault is best done at right angles to the fault plane (fig. 1.8A). However, the hazards associated with an unstable fault can often be avoided by drilling beneath the fault slope (B).

Other applications (fig. 1.9) include producing several reservoirs through a single wellbore (A), straightening an unintentionally deviated hole (B), and bypassing a stuck fish (C). A well drilled into the gas cap of a reservoir can be plugged back (D) and deviated to the oil zone to preserve gas drive pressure. Total production can be maximized by drilling a horizontal drain hole (E) to produce the zone evenly. A large area can be explored from a single wellbore (F).

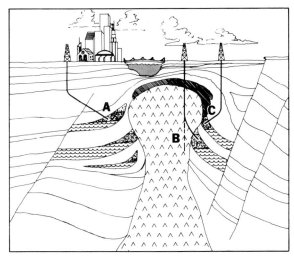

Figure 1.7. Directional wells drilled around geographic and geologic obstructions

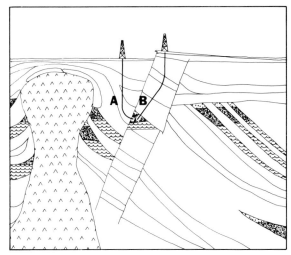

Figure 1.8. Directional drilling through and beneath a fault

The Well Plan

The operating company contracts with a directional drilling company to plan the hole. Provided with information on the type of well plan desired and the vertical depth of and horizontal distance to the target, the directional contractor uses a computer to draw vertical and horizontal projections showing how the well may be drilled at the lowest possible cost, with due consideration for safety and regulations. Among other factors that influence the final plan are –
1. formations to be drilled;
2. the hoisting, rotating, and pumping capacity of the rig;
3. mud and casing programs;
4. hole size; and
5. equipment availability.

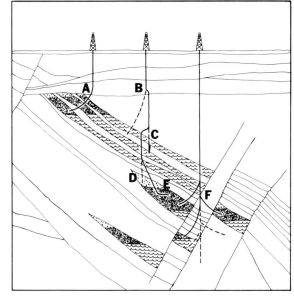

Figure 1.9. Miscellaneous applications of directional holes

Figure 1.10 shows a proposed Type I hole plan. The plan provides two views of the proposed well: a vertical projection, or *vertical section,* and a horizontal projection, or *plan view.*

The vertical section is a projection of the well on the plane that includes the surface location and the center of the target. The *target* or *objective* is usually a two- or three-dimensional zone rather than a point. *Deviation* (or *departure*) is the horizontal distance from the rotary table to the target; it is drawn to the same scale as depth. In figure 1.10, deviation is 2,691.8 feet (ft) and *true vertical depth* (TVD) is 8,750.0 ft.

Measured depth (MD), the length of the hole, is 9,273.1 ft. MD is always greater than TVD; the difference depends upon *drift angle* (or *inclination*), rate of angle buildup and dropoff, and, in the finished hole, unplanned deviations. *Course length* is the length of any portion of the hole, such as the amount of hole drilled between two directional survey stations. Total course length of the hole is its MD.

The plan view shows the proposed wellbore from above, with true (geographic, not magnetic) north at the top of the page. In the *quadrant* system, the horizontal direction of the target from the wellsite is expressed in terms of a number of degrees east or west of due north or due south. The well shown has a hole direction of S20°45'E, which means the target is located 20¾ degrees east of due south of the wellsite. (Hole direction can also be expressed as an *azimuth*–the number of degrees clockwise from due north. East is 90°; south is 180°. The azimuth corresponding to S20°45'E is 159.25°.) Departure is also shown on the horizontal projection; the scale may or may not be the same as that of the vertical projection.

quadrant

azimuth

The well in figure 1.10 is to be drilled vertically to 1,800 ft; here, at the *kickoff point* (KOP), the hole will be deviated 20°45' east of due south. The drift angle will be built up over a 900-ft course length (2,700 ft – 1,800 ft) to a final angle of 22°30' (22.5°). The average rate of angle buildup can be calculated by using the following formula:

$$\frac{100 \times (\text{final angle} - \text{initial angle})}{(\text{final MD} - \text{initial MD})}$$

The rate of angle buildup, or increase in drift angle, for this well is therefore

$$\frac{(100)(22.5 - 0.0)}{(2700 - 1800)}$$
$$= \frac{2250}{900}$$
$$= 2.5$$

or 2.5° per 100 ft of hole drilled. The final drift angle is maintained to target.

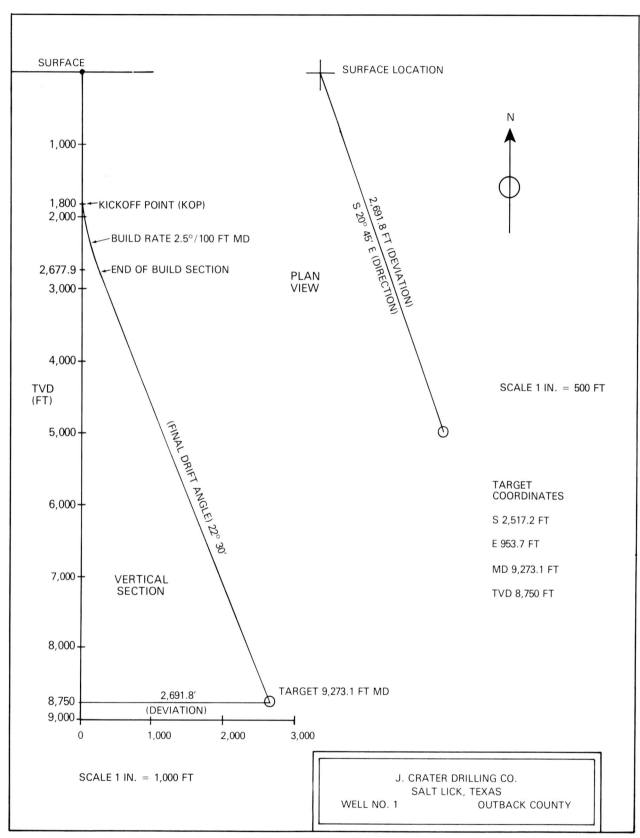

Figure 1.10. Plan for drilling a deviated well

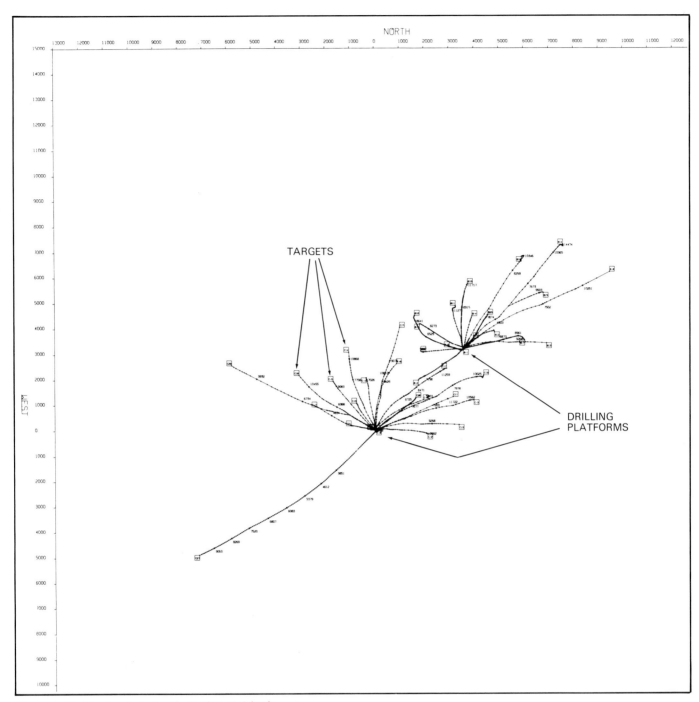

Figure 1.11. Horizontal structure plot at 1 inch = 1,000 feet (*Courtesy of Eastman Whipstock*)

Directional well plans are often generated by computer, especially in the case of multiple wells from an offshore drilling platform. For instance, the *horizontal structure plot* (fig. 1.11) shows plan views of wells already drilled radially out from two offshore platforms. The nearer portions of the wells on the northeast platform are shown at medium scale (fig. 1.12). The *horizontal slab* (fig 1.13) shows, at even larger scale, the planned trajectories between specified TVDs – in this case, the surface and 2,200 ft

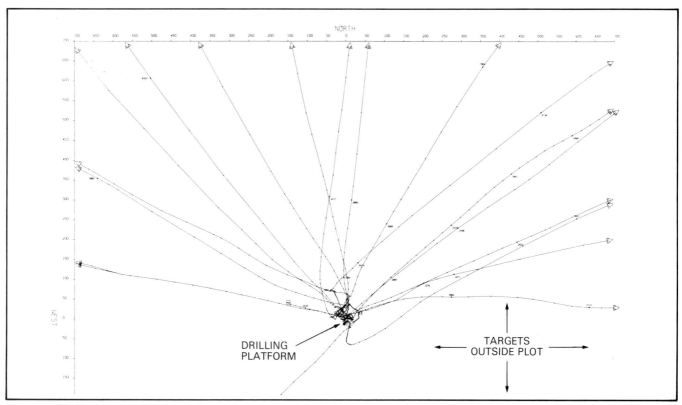
Figure 1.12. Horizontal structure plot at 1 inch = 50 feet (*Courtesy of Eastman Whipstock*)

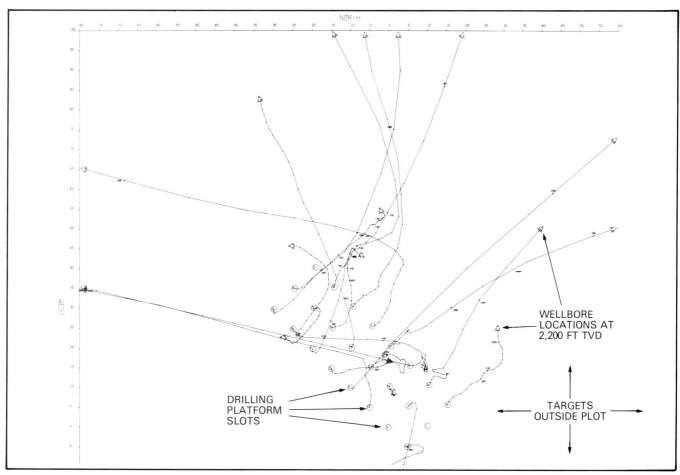
Figure 1.13. Horizontal slab plot at 1 inch = 5 feet (*Courtesy of Eastman Whipstock*)

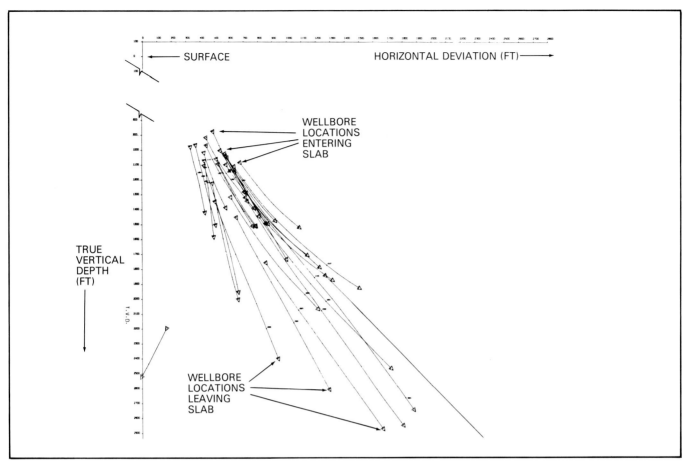

Figure 1.14. Vertical slab plot at 1 inch = 100 feet
(*Courtesy of Eastman Whipstock*)

TVD. The chart (fig. 1.14) shows portions of wells passing through a *vertical slab* of the earth located a given distance and direction from one of the drilling platforms. (The scales shown are those of the original charts, which have been reduced for this text.) These charts are especially useful for planning offshore wells so that they will not intersect existing wells; the computer can plot both the existing wells and the planned trajectory on the same chart.

DIRECTIONAL SURVEYING

Straight (nondirectional) holes are routinely measured to ensure that drift angle stays within specified limits. In directional drilling, however, both drift angle and direction must be determined at various depths to compare the actual course of the hole with the planned course (fig. 1.15). In general, two types of directional surveying are conducted during the drilling of a directional hole. One type is conducted by the directional supervisor, the other by a surveying service company.

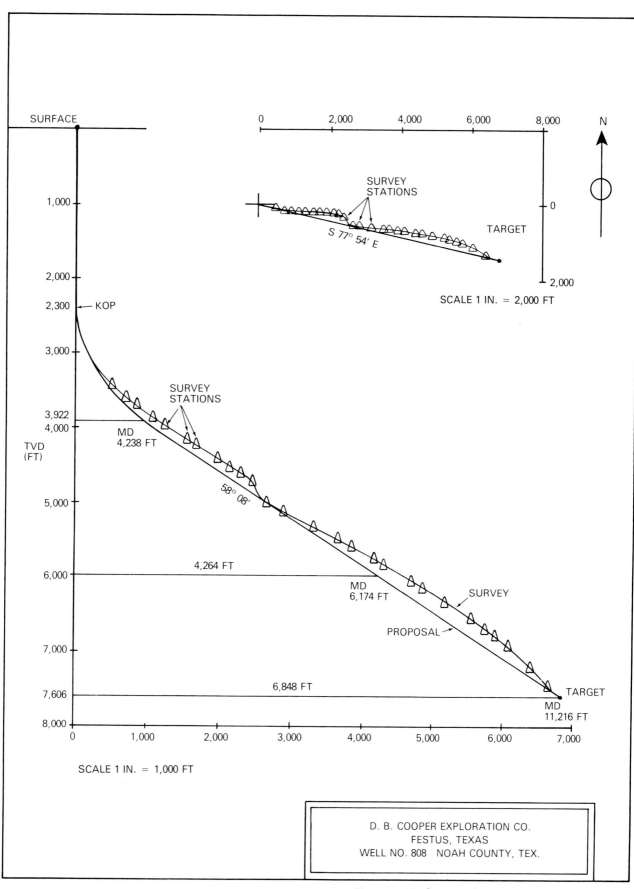

Figure 1.15. Survey plotted by computer on proposal

To monitor the progress of the hole, the directional supervisor conducts single-shot surveys whenever he likes. These surveys can often be completed, at small cost in rig time, during routine drilling operations—for instance, just before tripping out for a bit change. More advanced systems, such as steering tools or MWD (measurement while drilling) systems, furnish the directional supervisor with real-time directional data on the rig floor—that is, they show what is happening downhole during drilling.

The surveying service company, on the other hand, takes complete surveys of the hole as required by the operator— usually, either before or after setting a string of casing or upon completion of the hole. The instruments used are designed to take a series of readings at designated intervals. These readings are analyzed (usually by an office-based computer) to provide an accurate hole trajectory. This type of surveying program is separate from, and independent of, the directional supervisor's surveys.

Survey interval requirements vary among companies, states, and oil fields. Along the U. S. Gulf Coast, directional holes must be surveyed at least every 200 feet of hole drilled; on the Pacific coast, every 100 feet. Where wells are very close together, as on an offshore platform, a 30-foot survey interval may be required to avoid intersecting other wells. These wells are often kicked off with a steering tool (a directional survey instrument that shows, on a rig floor monitor, the inclination and direction of a downhole sensing unit).

A variety of directional surveying systems are in common use, ranging from older, simpler photographic instruments to the newest continuous-readout downhole monitoring systems. The best system for a particular situation depends upon the cost of using the system, the cost of rig time, and the degree of accuracy required.

Surveying with Photographic Instruments

The oldest type of directional instrument operates by making a downhole record that is brought to the surface to be analyzed. A photographic instrument is run into the hole and retrieved in one of three ways.
1. It may be run into and pulled out of the drill pipe on braided wireline.
2. It may be dropped free down the drill pipe, then retrieved with an overshot assembly run into the hole on a wireline.
3. It may be dropped free down the drill pipe when a trip becomes necessary (e.g., to change a dull bit), and recovered when the drill string is tripped out.

Figure 1.16, an exploded view of an instrument assembly, shows one of several different heads that may be made up on top, as well as shock absorber and bumper assemblies that reduce the shock of the downhole landing.

Magnetic single-shot. A typical magnetic survey instrument photographs a plumb bob against the background of a magnetic compass (fig. 1.17). The freely rotating compass aligns itself with the earth's magnetic field; the plumb bob hangs vertically. An electric light, focused on a single round disc of light-sensitive film, illuminates the pendulum and compass. In an inclined hole, the image of the plumb bob appears off-center in the photograph of the compass disc.

Conventional steel drill collars often become magnetized by the earth's magnetic field, causing magnetic compasses to give false readings. To obtain accurate magnetic survey readings, a magnetic instrument must be landed near the bit in a special drill collar made of a metal that is not easily magnetized, such as stainless steel. The survey must be taken far enough from the nearest magnetic metal to avoid magnetic distortions. The number of nonmagnetic collars and the location within them where the instrument must

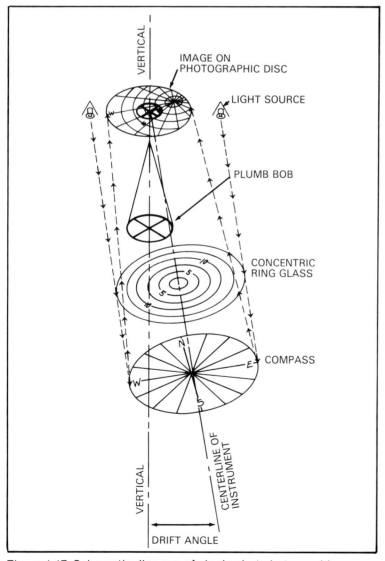

Figure 1.17. Schematic diagram of single-shot photographic survey instrument

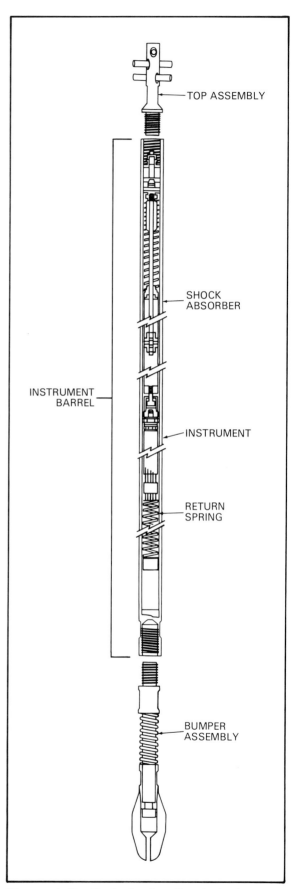

Figure 1.16. Typical directional instrument assembly

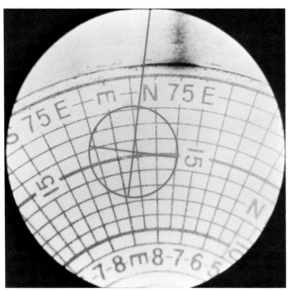

Figure 1.18. Magnetic single-shot survey disc (*Courtesy of Jordan Sawdo*)

magnetic reading

true reading

declination

land vary with magnetic latitude, drift angle, and hole direction. Since steel casing also becomes magnetized, magnetic surveys cannot be conducted in cased holes.

To make a directional survey, the photographic disc is inserted into the camera by using a light-tight injector device. A timer is made up in the assembly and set to allow time for the plumb bob and the compass card to stop moving before the survey reading is made. Often a motion sensor or a nonmagnetic sensor is included in the instrument assembly to activate the timer after it has stopped moving downhole and come to rest in the nonmagnetic collar. A nonmagnetic sensor is often used in shallow holes where a motion sensor might be affected by rig vibrations.

The instrument assembly is run downhole and seated in the nonmagnetic collar. The timer turns on the light, which shines through the plumb bob and is focused on the photographic disc. With the disc properly exposed, the light is switched off and the instrument retrieved. At the surface, the disc is developed and read.

Figure 1.18 shows a typical survey picture. The angle of the hole is directly related to how far off center the pendulum appears. In the example, the pendulum crosshairs fall on the 15° circle at a point 86° east of north of the compass center; the hole is going 15° N86°00'E at the survey point. This is a *magnetic reading*, because the instrument's compass points to the earth's magnetic poles. The north magnetic pole is several hundred miles from the north geographic pole, toward Hudson Bay. But directional hole plans, like standard maps and charts, are based on *true readings*, so the magnetic reading on the disc must be adjusted.

The difference between true and magnetic north readings is called *declination*. The declination depends upon where the survey is taken. Figure 1.19 is an *isogonic chart*—a map showing lines along which the declination is the same at all points. (Isogonic charts must be updated every 5 years or so because declination changes. The isogonic chart shown was drawn in 1975.) According to the chart, if the survey reading in figure 1.18 were taken near San Diego, California, true north would lie about 14°00' east of magnetic north. Thus 14°00' would be added to the magnetic reading in the northeast quadrant of the compass, giving a true reading of N100°00'E. However, quadrant readings are expressed as degrees east or west of due north or south and are always less than 90°. Since N90°00'E is the same as due east, N100°00'E is converted to a southeast reading: S80°00'E.

On the other hand, if the survey were taken near Raleigh, North Carolina, 5°30' would be subtracted, giving a true reading of N80°30'E. Notice that in Chattanooga, Tennessee, where declination is 0°, the magnetic reading would be the same as the true reading.

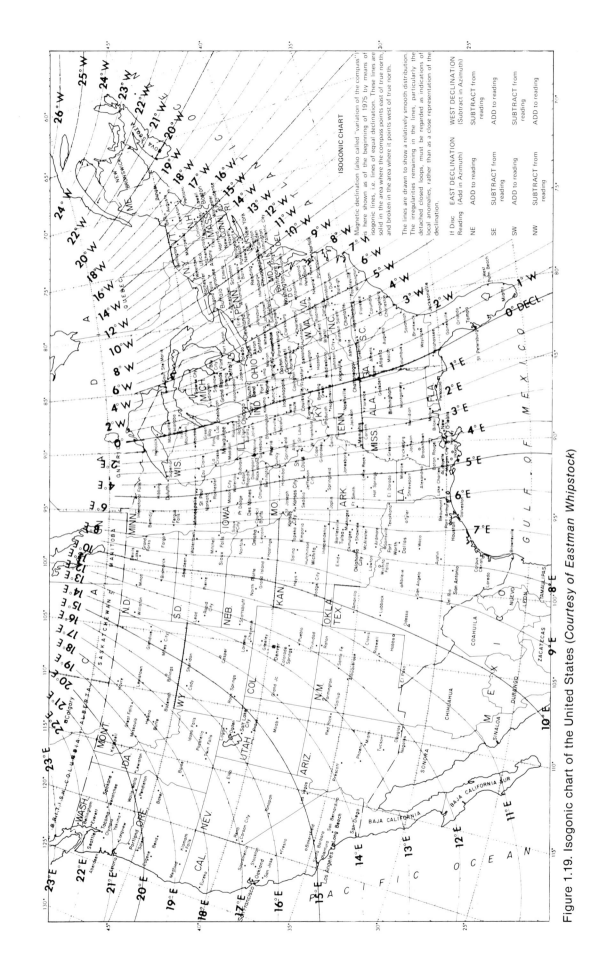

Figure 1.19. Isogonic chart of the United States (*Courtesy of Eastman Whipstock*)

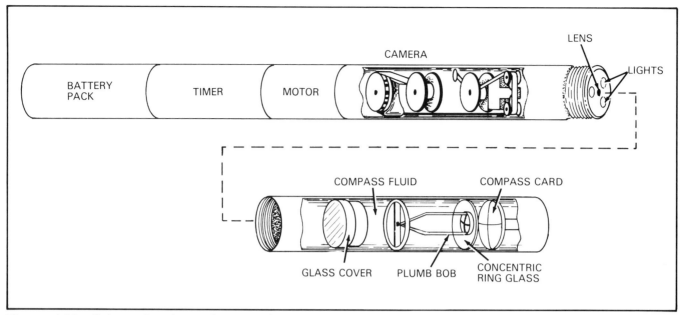

Figure 1.20. Schematic view of magnetic multishot instrument (*Courtesy of Eastman Whipstock*)

Magnetic multishot. The directional drilling plan calls for comprehensive surveys of the hole at certain times, such as just before setting a casing string. These surveys, usually conducted by a separate directional surveying contractor, are used to determine the overall trajectory of the hole in a single survey. The instruments used are called *multishot* instruments.

A *magnetic multishot* instrument works much like a magnetic single-shot; the main difference is that the multishot contains a built-in film-wound camera with a timer that automatically exposes and advances the film at preset intervals (fig. 1.20). Readings are analyzed and survey results plotted by the service company.

The magnetic multishot may be dropped free or lowered by wireline into the nonmagnetic collar. However, since a magnetic compass must stay within the nonmagnetic collar in order to function properly, the survey is taken as the instrument is brought to the surface by tripping out the drill string. Using a stopwatch, the surveyor keeps track of the depth at which the preset timer takes a shot. Only those shots taken at known depths with the drill string stationary are used in plotting the trajectory.

Gyroscopic multishot. Well casing, like conventional steel drill collars and pipe, becomes magnetized and disrupts magnetic compasses. Magnetic surveys thus become unreliable in cased wells or in open holes where nearby wells are cased, as in multiple wells drilled from an offshore platform. A *gyroscopic multishot* instrument may be used for directional surveys in or near cased wells or down the length of the drill string.

A *gyroscope* is a wheel or disc mounted to spin rapidly about one axis but free to rotate about one or both of two axes perpendicular to each other. The inertia of the spinning wheel tends to keep its axis pointed in one direction regardless of how the other axes are rotated. The *gyrocompass* in a gyroscopic multishot is a compass card linked to a gyroscope; the gyroscope itself is the weighted rotor of an electric motor spinning at 40,000 or so rpm. Unlike a magnetic compass, a gyrocompass is not affected by the earth's magnetic field. However, because gyroscopes are very sensitive to vibration and easily damaged by shock, they must be run into the drill string and retrieved by wireline. Gyro surveys must also be completed within a set time because gyros tend to drift gradually out of alignment; therefore, they are usually run going into the hole, rather than coming out (as with magnetic surveys).

Figure 1.21 shows the arrangement of working parts in a gyroscopic survey instrument. Before this instrument is run into the well, a directional pointer is aligned with a known point or direction (usually true north). The flywheel, or rotor, is driven at constant speed by an electric motor powered either by batteries or from the surface via wireline.

The instrument is wirelined down to the first drill collar above the bit. Like the magnetic multishot, it takes survey readings at preset intervals, but usually on the way down the drill string (fig. 1.22). The cross in each of the survey photographs is the image of the pendulum, indicating drift angle and direction. The directional pointer, attached to the gyro, appears as a long line extending outward from near the center.

Occasionally the directional drilling supervisor may wish to conduct a single-shot survey in cased hole or in open hole near cased wells. Instead of a magnetic single-shot, he can use a *gyroscopic single-shot* instrument. The gyro single-shot, like the gyro multishot, must be run and retrieved by wireline to avoid shock damage.

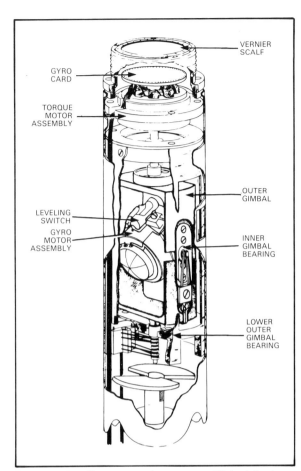

Figure 1.21. Gyroscope in gyro multishot instrument (*Courtesy of Eastman Whipstock*)

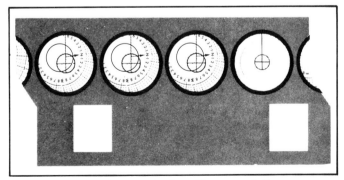

Figure 1.22. Gyro multishot film record (*Courtesy of NL Sperry-Sun*)

Surveying by Downhole Telemetry

The advantages of photographic surveying are its relative simplicity and its low cost. Disadvantages include the time required to run and retrieve the instrument or to start a survey all over again if the first survey is unusable. In drilling, time is money, and the longer the drill string sits in the hole, the greater is the chance of its getting stuck. An instrument that can provide the directional supervisor a real-time readout (one that shows what is happening at the moment the measurement is being taken), although initially more expensive, may save well costs in the long run.

telemetry

surface readout

Telemetry means measuring at a distance. A telemetry instrument having a measuring unit downhole and a monitor at the surface is commonly known as a *surface-readout* instrument. Since the directional supervisor cannot see the downhole sensing unit, the directional data it records must be converted into signal pulses and transmitted uphole to the monitor. Some instruments transmit these signals via conducting wireline; others, through the drilling fluid. On the rig floor, small computers, or *microprocessors,* convert the data to readable form and display them on a dial or an LED (light-emitting diode) display (fig. 1.23) or sometimes on a tabular or graphic printout. Rig floor readouts generally display azimuth (0°–360° directional readings) rather than quadrant directions.

microprocessor

Telemetering instruments, in general, sense hole direction in one of two ways: with gyroscopes or with magnetometers. Gyroscopes are sensitive to vibration and easily disrupted or damaged by shock. Instruments containing gyros are run downhole only with drilling operations suspended, and pulled out again before drilling resumes. *Magnetometers* (electromagnetic devices that sense the intensity and direction of the earth's magnetic field) are comparatively rugged, can be left in the drill string while the bit is turning, and so are commonly used in measurement while drilling (MWD) systems, which include steering tools and mud-pulse systems (described later).

Figure 1.23. Surface readout for downhole telemetry instrument (*Courtesy of Eastman Whipstock*)

Gyroscopic telemetry. Because gyros are sensitive to vibration and shock, the drill string must be at rest while any gyro instrument is run downhole and retrieved. Therefore, a gyro telemetry instrument is used much like a gyro photographic instrument. The difference lies in how the readings are obtained. In a photographic system, directional data are not available (nor the validity of the results known) until the film is developed and analyzed. With gyroscopic telemetry, however, the reading is displayed on a rig floor monitor as soon as the instrument reaches the downhole survey location. The survey is known to be valid before the instrument is retrieved, and drilling can proceed without delay.

A simple gyroscopic telemetry instrument contains a single gyroscope and an *accelerometer* (a device that senses changes in motion) to measure direction and inclination. As with a photographic instrument, the gyro is aligned with a surface reference point or direction and the instrument is run downhole to the survey point. Instead of producing photographs, however, the telemetering instrument reads direction and inclination electronically and transmits the signals up the wireline to the surface.

One disadvantage of a gyroscope is its tendency to drift off its initial heading. Friction and other forces slow it down and cause it to rotate away from its original reference point. The longer the gyro is downhole, the larger the probable error becomes. Largely as a result of space-age missile technology, new types of gyroscopes have been developed to overcome this problem and increase the accuracy and efficiency of gyro surveying.

True north reference surveying. A *rate gyro* is a gyroscope that is not allowed to gimbal (pivot) freely but is held in position as it spins within the instrument barrel (fig. 1.24). A rate gyro does not drift because it is fixed with reference to the instrument case; it does not require alignment with a fixed reference point. Used in combination with accelerometers, a rate gyro senses the difference in direction between the borehole and the earth's rotational axis. For this reason, a rate gyro is sometimes called a north-seeking gyro, and surveying with a rate gyro is called *true north reference surveying*.

A rate gyro measures hole inclination and direction independently at each survey station. With conventional gyros, a power stoppage means the instrument must be retrieved to realign the gyro, and the survey has to be rerun from the surface. With a rate gyro, the survey can be restarted where it left off because the gyro does not have to be realigned.

Current rate-gyro instruments are not well suited for drilling in the far north because there is not enough earth rotation at high latitudes for the rate gyro to measure. Rate gyros are even more sensitive to vibrations than conventional gyros and therefore harder to use offshore.

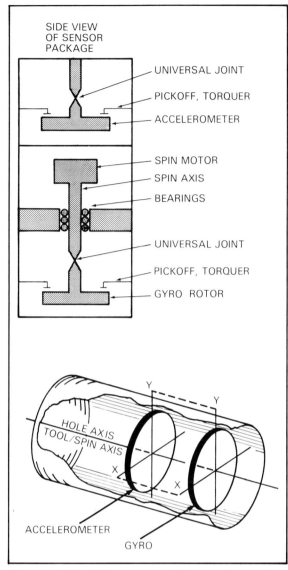

Figure 1.24. Rate gyro with accelerometer (*Courtesy of Oil and Gas Journal, April 11, 1983*)

21

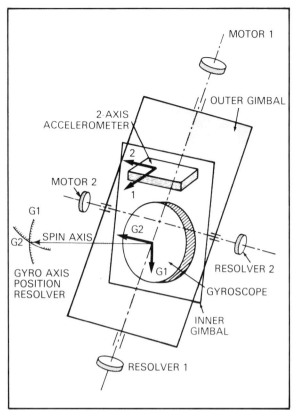

Figure 1.25. Schematic diagram of continuous guidance tool (*Courtesy of Society of Petroleum Engineers*)

With many true north reference surveying systems, the instrument has to come to rest to obtain readings. A *continuous guidance tool,* however, as its name implies, can read and transmit directional data while moving downhole or uphole (fig. 1.25).

Inertial measurement. A multiple-gyro aerospace guidance device called an *inertial platform* measures directional data in an *inertial measurement system.* The inertial platform is a cluster of three gyros and three accelerometers that is free to rotate in any direction (fig. 1.26). The gyros keep the inertial platform oriented vertically along the meridian (longitude) of the well location. The accelerometers measure the total movement in all three dimensions; microprocessors and the surface computer convert these data into three-dimensional coordinates for each survey station. More than 10 inches in diameter, this instrument cannot be run inside the drill string like other gyro instruments, but must be run on cable or drill pipe in open or cased hole. Gyro drift is measured each time the instrument comes to a stop; the computer takes this into account, maintaining accuracy without manual recalibration.

Since gyroscopes cannot be subjected to excessive vibration or shock, drilling must be suspended for up to several hours in order to run and retrieve any gyroscopic instrument. However, there are two nongyroscopic telemetry

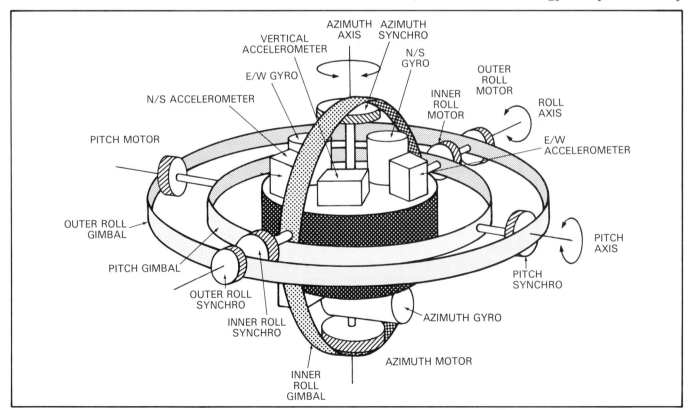

Figure 1.26. Schematic view of inertial platform (*Courtesy of Pétrole Informations, no. 1561, January, 1982*)

systems that can be left in the hole for *measurement while drilling* (MWD) – that is, directional surveying during drilling operations. One MWD system transmits data to the surface via wireline; the other, through the drilling fluid.

Steering tool. A *steering tool* is a wireline telemetry surveying instrument that measures inclination and direction while drilling is in progress (fig. 1.27). Because of the wireline, a steering tool can be used only with a *downhole motor* – a drill string motor that turns the bit while the drill string remains fixed.

A steering tool contains magnetometers that continuously measure hole direction and inclination and tool face orientation. Signals from the magnetometers are transmitted uphole via wireline to a rig floor computer, which converts the signals and displays directional data on the readout. The instrument enables the operator both to survey and to orient a downhole motor while actually using a deflection tool to make hole.

(A deflection tool changes the course of the hole by deflecting the bit to one side. That side of the tool is called the *tool face*. The direction the tool face is turned, its *orientation*, directly affects the direction the hole is curved. Deflection tools, downhole motors, and orientation will be discussed in later sections.)

Mud-pulse telemetry. Another type of directional telemetry instrument transmits signals uphole through the drilling fluid, allowing the driller to obtain real-time directional and other data without a wireline and, therefore, while making hole with a rotating drill string. Like other telemetry systems, a *mud-pulse telemetry* system has two main components: a downhole unit that senses direction and inclination, and a rig floor unit that displays the data. Microprocessors and a transmitter in the downhole unit convert survey data into a series of pressure pulses. Positive mud pulses are a series of pressure increases; negative pulses are pressure decreases. Signals can also be transmitted on a continuous or carrier wave, like radio signals. A computer at the surface interprets the signals and feeds them to the readout.

The term *measurement while drilling (MWD)* is often used as a synonym for *mud-pulse telemetry;* however, *MWD* is also used in a more general sense to mean any system of measuring downhole conditions during routine drilling operations.

A positive mud-pulse system in common use measures both drift angle and direction and transmits the information to the surface coded in a binary signal. The downhole unit, mounted inside a nonmagnetic collar, contains

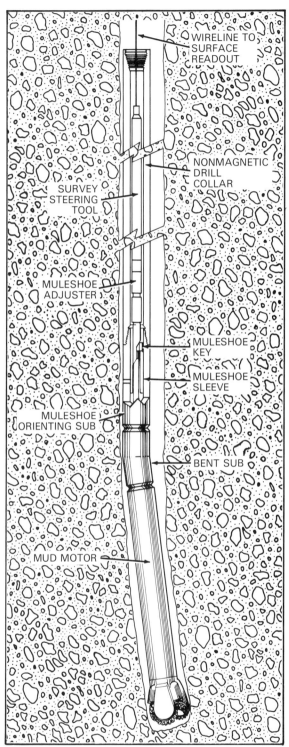

Figure 1.27. Steering tool downhole assembly (*Courtesy of NL Sperry-Sun*)

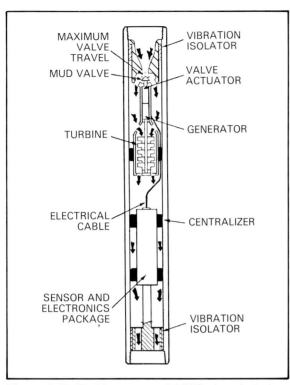

magnetometers and accelerometers for measuring inclination and direction. Circulating drilling fluid spins a turbine to power the transmitter (fig. 1.28).

This system can be used in either a rotating or a nonrotating drill string. In a rotating string, a rotation detector is included in the assembly to trigger angle measurement when rotation is stopped. Full circulation is maintained to power the transmitter. With a downhole motor, drilling does not have to be interrupted; direction, inclination, and tool face orientation are measured continuously and transmitted on demand as long as circulation is maintained.

Another turbine-powered system contains three accelerometers and three magnetometers and transmits negative pulses, which can carry more data per second than positive pulses. Measurements are made when rotation is stopped. Data are displayed on a compass and digital readout, and a printer provides a hard copy of survey data for later analysis (fig. 1.29). Other negative-pulse systems are powered by batteries.

Figure 1.28. Turbine-powered MWD downhole instrument (*Courtesy of Oil and Gas Journal*)

Figure 1.29. Negative mud-pulse MWD system (*Courtesy of Gearhart Industries*)

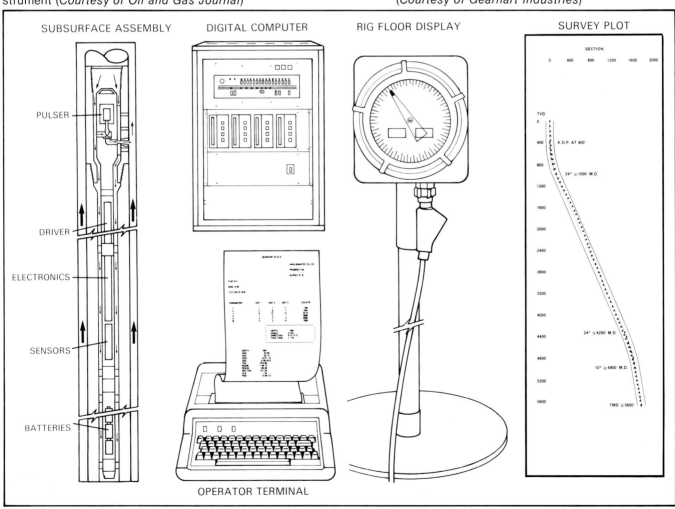

Some downhole telemetry devices provide more than directional data. One mud-pulse system measures azimuth, inclination, tool face orientation, weight on bit, torque, formation radioactivity, formation resistivity, and downhole temperature. Data are transmitted to the surface on a continuous or carrier wave.

Not all mud-pulse systems require an electric power supply. One fully mechanical device uses a spring-loaded mechanism to measure inclination. Shutting down circulation allows the mechanism to set itself (fig. 1.30). Restarting the mud pump then causes the device to send a series of pressure pulses up through the mud inside the drill stem. At the surface, these pulses are displayed on a strip-chart recorder (fig. 1.31). The number of pulses is directly proportional to the drift angle; the range of measurement can be varied. This inclination-measuring sub can be used in rotating or nonrotating systems. A separate, direction-measuring sub, consisting of a magnetic compass inside a nonmagnetic collar, is used only with a downhole motor.

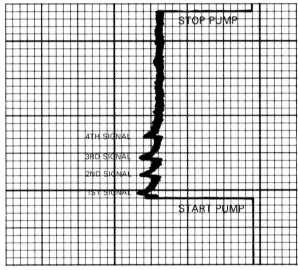

Figure 1.31. Readout for pressure-pulse directional instrument (*Courtesy of Sii Dyna-Drill*)

Plotting Survey Results

Often the directional supervisor does not have to concern himself with plotting the results of directional surveys. Surveying contractors who conduct multishot surveys provide the directional supervisor with a complete plot of the well trajectory. Directional data from many of the newer telemetry systems can be fed directly (or from tape) into computers that calculate and plot a well trajectory automatically. A computer-generated trajectory can be superimposed on the well plan (fig. 1.29). The directional supervisor's job is then simplified; he advises the driller on how to keep the trajectory as close as possible to the plan.

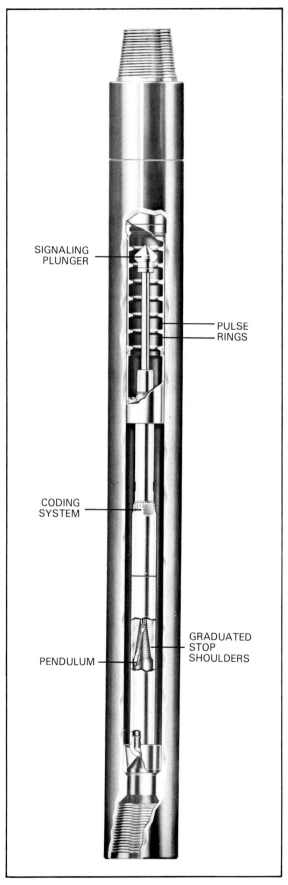

Figure 1.30. Mechanical mud-pulse drift-angle measuring system (*Courtesy of Sii Dyna-Drill*)

The three-dimensional coordinates of a survey station are expressed in terms of true vertical depth, distance north or south of the well site, and distance east or west of the well site. On the sample printout shown in figure 1.32, the last survey station is located at a TVD of 3,251.06 ft, 917.47 ft S and 651.13 ft E of the well site.

When conducting his own surveys, the directional supervisor usually calculates and plots the coordinates of each survey station himself. To do this, he must know four things:
1. Drift angle
2. Direction
3. Course length from the last survey station
4. Coordinates of the last survey station

Each point is plotted relative to the one above. An error at any point will throw off all subsequent points by the same amount. The error accumulates—hence the term *cumulative error*. To check for cumulative error when a gyroscopic multishot instrument is used, a few stations are

cumulative error

MEASURED DEPTH (FEET)	DRIFT ANGLE D	DRIFT ANGLE M	DRIFT DIRECTION D	COURSE LENGTH (FEET)	TRUE VERTICAL DEPTH (FEET)	RECTANGULAR COORDINATES (FEET)		DOGLEG SEVERITY (DG/100 FT)
0	0	0	0	0	0.00	0.00	0.00	0.00
951	0	0	0	951	951.00	0.00	0.00	0.00
1059	1	30	S 48 E	108	1058.99	0.95 S	1.05 E	1.39
1151	5	0	S 37 E	92	1150.83	4.78 S	4.57 E	3.85
1244	9	30	S 39 E	93	1243.06	14.03 S	11.79 E	4.85
1337	13	0	S 42 E	93	1334.26	27.82 S	23.57 E	3.81
1429	17	0	S 41 E	92	1423.10	45.65 S	39.35 E	4.36
1522	17	45	S 43 E	93	1511.86	66.29 S	57.93 E	1.03
1615	19	0	S 39 E	93	1600.12	88.41 S	77.16 E	1.91
1707	21	0	S 42 E	92	1686.56	112.33 S	97.59 E	2.44
1800	22	15	S 41 E	93	1773.02	138.00 S	120.30 E	1.40
1893	23	30	S 40 E	93	1858.70	165.49 S	143.78 E	1.41
1985	24	30	S 39 E	92	1942.75	194.36 S	167.58 E	1.17
2078	26	0	S 40 E	93	2026.86	224.97 S	192.81 E	1.68
2171	26	0	S 35 E	93	2109.72	258.46 S	218.50 E	3.25
2263	29	0	S 40 E	92	2190.57	293.27 S	245.22 E	2.81
2356	30	30	S 35 E	93	2271.31	329.87 S	273.30 E	3.12
2449	31	30	S 34 E	93	2351.02	369.35 S	300.43 E	1.21
2542	33	0	S 34 E	93	2429.67	410.49 S	328.18 E	1.61
2634	33	45	S 33 E	92	2506.50	452.69 S	358.12 E	1.01
2727	33	15	S 32 E	93	2584.05	495.98 S	383.70 E	0.80
2820	33	30	S 32 E	93	2661.72	539.37 S	410.81 E	0.27
2912	33	45	S 35 E	92	2738.32	581.85 S	438.92 E	1.83
3005	35	0	S 36 E	93	2815.08	624.60 S	469.41 E	1.47
3098	36	30	S 31 E	93	2890.55	669.89 S	499.39 E	3.53
3190	39	0	S 31 E	92	2963.29	718.17 S	528.40 E	2.72
3283	40	15	S 32 E	93	3034.92	768.74 S	559.39 E	1.51
3376	39	45	S 31 E	93	3106.16	819.70 S	590.62 E	0.88
3468	38	45	S 33 E	92	3177.41	869.07 S	621.47 E	1.75
3581	36	30	S 30 E	93	3251.06	917.47 S	651.13 E	3.12

Figure 1.32. A printout generated by a survey program (*Courtesy of Eastman Whipstock*)

surveyed coming out of the hole. The survey should *close* – that is, the coordinates of the final point (the surface) should be very nearly the same as those of the starting point (fig. 1.33).

Surveys conducted by using continuous guidance or inertial measurement systems provide coordinates for each survey station relative to the starting point or surface location, eliminating the problem of cumulative error, and are most practical for wells for which great accuracy is required, such as those on a multiwell offshore platform.

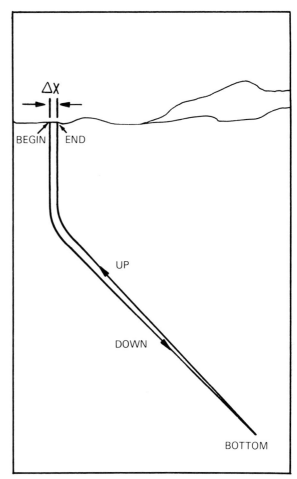

Figure 1.33. Closure

LESSON 1 QUESTIONS

Put the letter for the best answer in the blank before each question.
Questions 1–22 cover material in **Introduction** and **Directional Wells.**

_____ 1. The first use of controlled directional drilling was to –
  A. control a blowout.
  B. drill through a salt dome.
  C. reach an offshore reservoir.
  D. sidetrack a stuck fish.

_____ 2. The directional drilling supervisor's job does *not* include –
  A. conducting single-shot directional surveys.
  B. advising the driller on selection of directional drilling equipment.
  C. testing well fluids.
  D. helping the driller orient deflection tools.

_____ 3. Which of the drawings below represents a Type II directional well?

A   B   C   D

_____ 4. The relief well shown in figure 1.6 is –
  A. a Type I well.
  B. a Type II well.
  C. a Type III well.
  D. a straight hole.

_____ 5. In the directional well plan, hole direction is –
  A. shown in the plan view or horizontal projection.
  B. expressed as a magnetic direction.
  C. the same as hole inclination.
  D. the total course length of the hole.

Match the descriptions on the right with the terms on the left by placing the letters in the appropriate blanks. Letters may be used more than once.

_____ 6. Departure

_____ 7. Course length

_____ 8. TVD

_____ 9. MD

_____ 10. Inclination

_____ 11. Horizontal projection

_____ 12. Target

_____ 13. Azimuth

_____ 14. KOP

_____ 15. Plan view

_____ 16. Vertical section

_____ 17. Objective

_____ 18. Deviation

_____ 19. Drift angle

A. Underground location where well is supposed to bottom out
B. Beginning of deflection
C. Difference in elevation between bottom and top of hole
D. Side view of well plan
E. Overhead view of well plan
F. Horizontal distance from top of well to target
G. Angle of wellbore off vertical
H. Length of any part of hole
I. Total length of hole drilled
J. Degrees clockwise from due north

_____ 20. The measured drift angle of a hole is 10°30' at 1,225 ft MD and 13°30' at 1,300 ft MD. Average rate of drift angle change through this section is—
A. 3°/100 ft.
B. 4°/100 ft.
C. 12°/100 ft.
D. none of the above values.

_____ 21. Drift angle is 5½° at 850 ft MD. If the driller builds angle at 2½°/100 ft, drift angle at 1,100 ft MD will be—
A. 3°.
B. 6°15'.
C. 11°45'.
D. 13°45'.

_____ 22. The well plan is influenced by—
A. formations to be drilled.
B. surface location and target location.
C. the hoisting capacity of the rig.
D. the casing program.
E. all of the above factors.

Look again at the section titled **Directional Surveying.** Answer questions 23–40.

_____ 23. Multishot directional surveys are usually conducted by –
  A. the driller.
  B. the directional drilling supervisor.
  C. the operating company.
  D. a directional surveying company.

_____ 24. A directional surveying instrument measures –
  A. drift angle and direction.
  B. drift angle and MD.
  C. direction and TVD.
  D. rate of angle buildup and course length.

_____ 25. Magnetic directional surveys can be conducted –
  A. in uncased hole, using nonmagnetic collars.
  B. in cased hole, using nonmagnetic collars.
  C. in cased hole, using conventional steel collars.
  D. in uncased hole, using conventional steel collars.

_____ 26. Factors determining the number of nonmagnetic collars needed in the downhole assembly include –
  A. gyroscope rotation speed and direction.
  B. magnetic latitude and drift angle.
  C. type of photographic film used.
  D. temperature and time of day.
  E. all of the above.

_____ 27. A motion sensor is used in a directional instrument assembly to –
  A. take readings during trips in and out of the hole.
  B. prevent taking readings until the instrument comes to rest.
  C. take survey readings inside a rotating drill string.
  D. save money.

_____ 28. The photographic disc at right is read as follows:
  A. drift angle 6°, direction N135°00'E.
  B. drift angle S45°00'E, MD 600 ft.
  C. TVD 6,000 ft, direction S45°00'E.
  D. drift angle 6°, direction S45°00'E.
  E. none of the above.

31

_____ 29. The difference between true and magnetic north is—
   A. always 15° or more.
   B. negligible in directional surveying.
   C. called *declination*.
   D. never 0°.

_____ 30. A magnetic compass reading of N52°00'W taken near Albuquerque, New Mexico, in 1975 would represent a true reading of—
   A. N64°00'W.
   B. N64°00'E.
   C. S40°00'W.
   D. N40°00'W.

_____ 31. A gyroscopic multishot instrument measures—
   A. true direction and inclination.
   B. magnetic direction and TVD.
   C. magnetic direction and inclination.
   D. hole diameter.

_____ 32. A gyroscopic single-shot instrument is—
   A. dropped free downhole to the bit.
   B. wirelined to the bottom of the drill string.
   C. run in open hole and left there during drilling.
   D. retrieved by tripping out the drill string.

_____ 33. MWD means—
   A. measured width and deviation.
   B. median well depth.
   C. measurement while drilling.
   D. mud weight density.

_____ 34. An accelerometer is a device that—
   A. pivots to seek geographic north.
   B. speeds up the drilling process.
   C. measures changes in motion.
   D. measures magnetic direction.

_____ 35. A rate gyro must be corrected for drift. (T/F)

_____ 36. An inertial measurement system has—
   A. a magnetic compass.
   B. gyros and accelerometers.
   C. magnetometers.
   D. a mud turbine.

_____ 37. Steering tools are used to –
   A. photograph the hole while the bit is being changed.
   B. show direction, inclination, and tool face orientation in making hole.
   C. drill out a float shoe after a casing string is cemented.
   D. turn the rig when it is being moved.

_____ 38. Positive mud-pressure pulses can transmit data faster than negative pulses. (T/F)

_____ 39. The three-dimensional coordinates of a survey station might be expressed as follows:
   A. TVD 1,280.5 ft, 86.8 ft N, 128.5 ft E
   B. TVD 1,280.5 ft, 8.5° S25°45'W
   C. MD 1,360 ft, 10.3° NW
   D. 19°, 2,800 ft W, 3.75°N

_____ 40. Surveys using inertial measurement or continuous guidance systems –
   A. relate each survey station directly to the surface location.
   B. are subject to cumulative error.
   C. cannot be conducted on offshore platforms.
   D. are less accurate than photographic surveys.

Lesson 2
Controlled Directional Drilling
(Part II)

Changing the Course of the Hole

Special Problems in Directional Drilling

Summary

Lesson 2
CONTROLLED DIRECTIONAL DRILLING, Part II

CHANGING THE COURSE OF THE HOLE

A basic requirement in drilling a directional well is some means of changing the course of the hole. Generally, a driller either uses a specially-designed deflection tool or modifies the bottomhole assembly he is using to drill ahead.

Deflection Tools

A *deflection tool* is a drill string device that causes the bit to drill at an angle to the existing hole. Deflection tools are sometimes called *kickoff tools* because they are used at the kickoff point (KOP) to start building angle. There are many different types of deflection tools, ranging from the primitive (but rugged) whipstock to the state-of-the-art downhole motor. The directional supervisor's choice depends upon the degree of deflection needed, formation hardness, hole depth, temperature, presence or absence of casing, and economics. The most important factor is the formation in which the deflection is to be made, because it is the only factor beyond human control.

Deflection tools cause the bit to drill in a preferred direction because of the way the tool is designed or made up in the drill string. The *tool face* is the direction in which the bit tends to drill. It is usually marked with a scribe line. The tool face must be *oriented,* or turned in a particular direction, to deflect the hole as desired. It is important to remember that the direction the tool is faced is not necessarily the same as the desired hole direction.

Whipstocks. The earliest deflection tool was a tapered slab of wood placed in the bottom of the hole to force the bit toward one side. It was called a *whipstock* because it resembled the handle of a whip. A typical modern version, such as the casing whipstock shown in figure 2.1, is made of steel and has a tapered concave groove (the tool face) to guide the rotating bit against the casing. It builds hole angle 2° to 3° over its 6-ft to 12-ft length.

A variety of whipstocks is available for special purposes. A nonretrievable whipstock, for instance, may be used to bypass a stuck fish; it is left in place after the deflection has been accomplished. A retrievable whipstock, on the other hand, is tripped out with the bit. A circulating whipstock directs fluid to the bottom of the hole to flush out cuttings and ensure a clean seat for the tool.

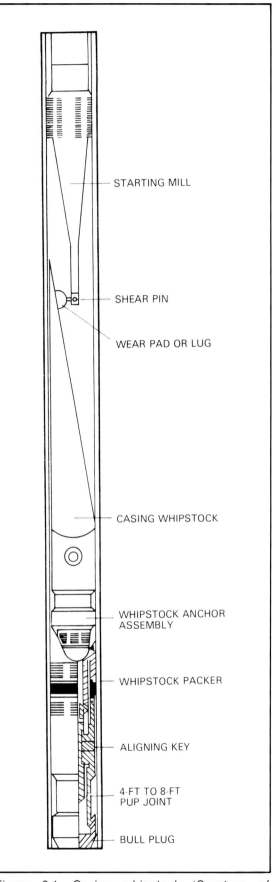

Figure 2.1. Casing whipstock (*Courtesy of Eastman Whipstock*)

37

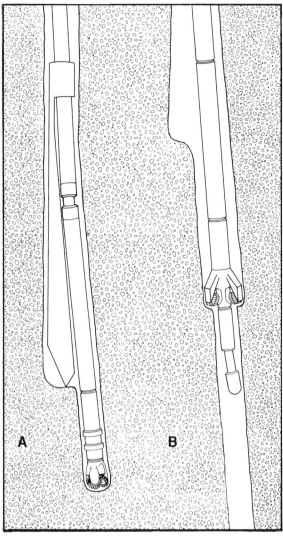

Figure 2.2. Drilling operation with whipstock: *A*, drilling off; *B*, follow-up reaming

One drawback of a whipstock is that it drills an under-gauge hole that must be reamed out in a separate operation (fig. 2.2), requiring more trip time. It may turn in the hole; several orientation surveys may be needed to determine whether it is seated properly. Another disadvantage is that only 15 to 20 feet of hole can be drilled at a time. As a result, whipstocks have been largely supplanted by more sophisticated directional tools and are now used only in unusual situations.

Jet deflection bits. Where formations are relatively soft, a *jet deflection bit* can be used to deviate the hole. A conventional roller cone bit is modified by equipping it with one oversize nozzle and closing off or reducing others, or by replacing a roller cone with a large nozzle (fig. 2.3). The bit is run into the hole on an angle-building bottomhole assembly (discussed later). The tool face (the side of the bit with the oversize nozzle) is oriented in the desired direction, the pumps started, and the drill string worked slowly up and down, without rotation, about 10 feet off bottom. This action washes out the formation on one side (fig. 2.4). When rotation is started and weight applied, the bit tends to follow the path of least resistance – the washed-out section. Extra weight is applied to bow the drill collars, and drilling continues until the correct hole angle is attained.

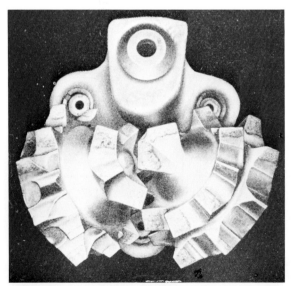

Figure 2.3. Jet deflection roller cone bit

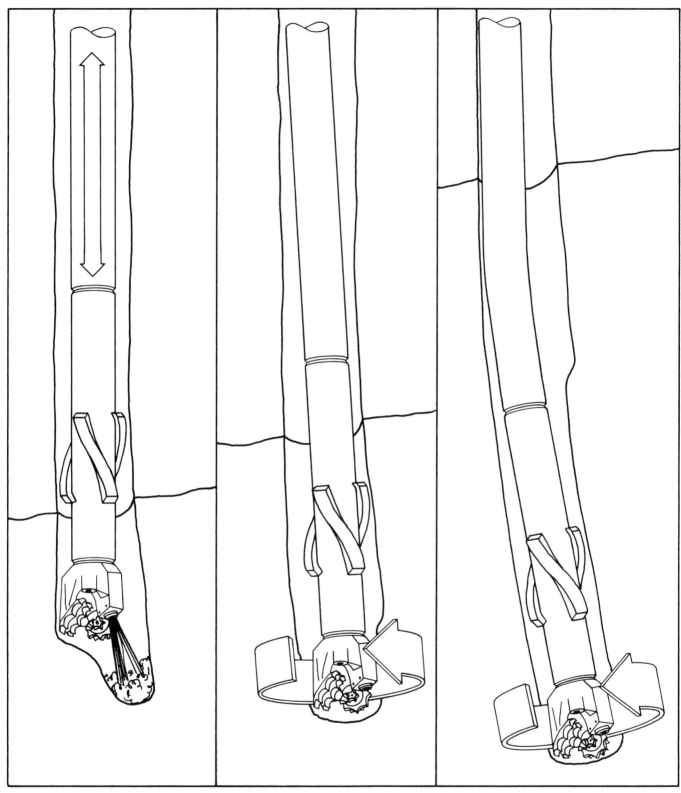

Figure 2.4. Deflecting hole with jet deflection bit

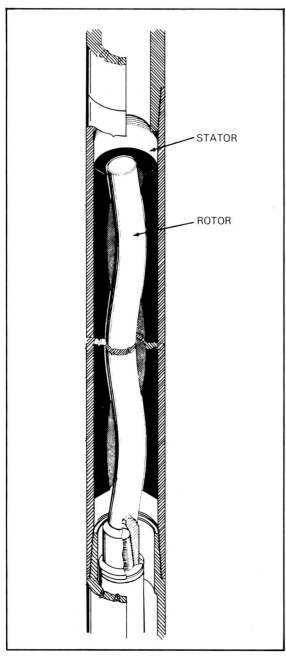

Figure 2.5. Cutaway view of positive-displacement motor (*Courtesy of Sii Dyna-Drill*)

Downhole motors. The most commonly used deflection tool—the *downhole motor,* or *mud motor*—has several advantages over older types of deflection tools. A downhole motor drills full-gauge hole, so follow-up reaming or reboring is not needed to open the deviated hole to full gauge. It can be used to make multiple deviations without coming out of the hole, to make course corrections after the well has been kicked off, to drill through bridges (obstructions), and to clean out bottomhole cuttings before beginning a deviation. It operates without drill string rotation and so reduces rig maintenance. It also drills efficiently at high speed, 300–1,000 rpm, compared to 50–150 rpm with conventional rotation.

The *positive-displacement motor,* one type of mud motor, is powered by the flow of drilling fluid (mud, water, or air) down the drill string. Its two basic parts are a stationary *stator* and a rotating *rotor* (fig. 2.5). Drilling fluid is pumped downward between the rubber-lined spiral stator and the wave-shaped rotor, forcing the rotor to turn and to transmit the power of the flowing mud to the bit. A dump valve opens when there is no circulating pressure in the drill string, allowing drilling fluid to bypass the motor and fill or drain from the drill string during tripping. When the pump is started, pump pressure closes the valve to direct fluid through the motor.

After the motor has been run into the hole and started, the bit is set on bottom. In a positive-displacement motor, drilling torque is proportional to the pressure loss through the tool; in other words, the higher the pressure loss, the greater is the torque. Conversely, circulating pressure increases as more weight is applied to the bit; excessive weight will stall the motor, so drilling with this tool requires coordinating available pump pressure and weight on bit. The drill string is not allowed to rotate while the deviated section is being drilled but may be rotated slowly when straight hole is being drilled.

The other principal type of mud motor, the *downhole turbine,* operates only with a liquid drilling fluid, such as mud or salt water. The downhole turbine motor contains bladed rotors and stators (fig. 2.6). The stators are attached to the outer case, the rotors to the drive shaft. Each rotor-stator pair is called a *stage;* a typical motor has 75 or more stages. The stators direct the flow of drilling mud onto the rotor blades, forcing the rotors and drive shaft to rotate to the right.

A standard downhole motor is not, by itself, a deflection tool. To deviate the hole, a *bent sub* is made up in the drill string between the drill collars and the downhole motor (fig. 2.7). A bent sub is a short collar with its pin and box

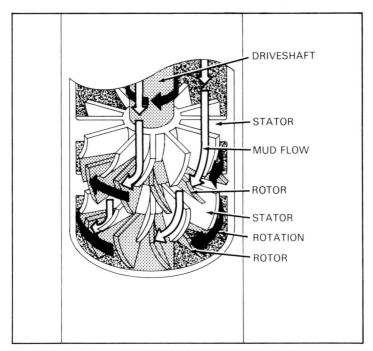

Figure 2.6. Cutaway view of turbine motor

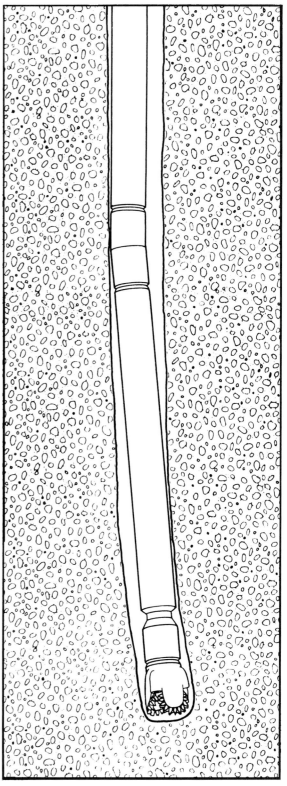

Figure 2.7. Positive-displacement motor deflected with bent sub

threads cut concentric with an axis bent 1° to 2½° offline (fig. 2.8). The tool face of the downhole motor assembly is the direction the sub is bent, usually marked with a scribe line for precise alignment. A flexible joint (fig. 2.9), which is run into the hole straight and kicked off at an angle

Figure 2.8. Cutaway view of bent sub (*Courtesy of Wilson Industries*)

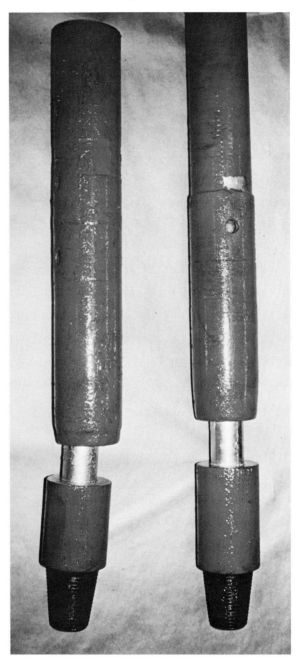

Figure 2.9. Flexible joint

downhole, is sometimes used in place of a bent sub. A positive-displacement motor can also be built with a housing bent 1° to 2° offline (fig. 2.10).

Downhole motors, unlike other deflection tools, are subject to *reactive torque* – the tendency of the drill string to turn in a direction opposite that of the bit. As the stator deflects drilling fluid to turn the rotor to the right, the stator itself is subjected to a left-turning force. Depending upon the length of the drill string and the type of formation being drilled, the drill pipe will twist or wind up when power is applied to the bit, causing the tool to drill toward the wrong direction. The driller learns by experience how much to compensate for reactive torque. A common rule of thumb is to allow 10° per 1,000 ft in drilling soft formations and 5° per 1,000 ft in hard formations. In other words, the motor is faced 5° or 10° to the right of the desired direction per 1,000 ft MD. Then, when the motor is started and weight applied to the bit, reactive torque will turn the tool back to the desired direction.

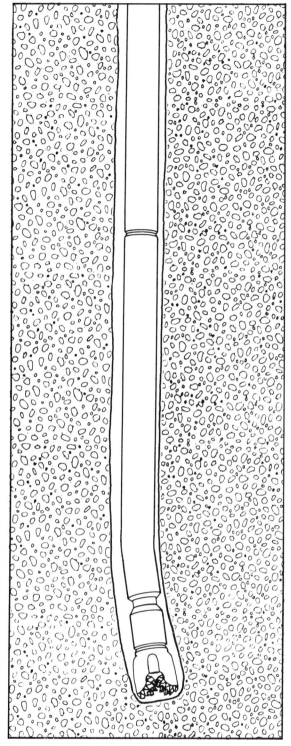

Figure 2.10. Positive-displacement motor with bent housing

Figure 2.11. Schematic diagram of downhole motor with steering tool (*Courtesy of NL Sperry-Sun*)

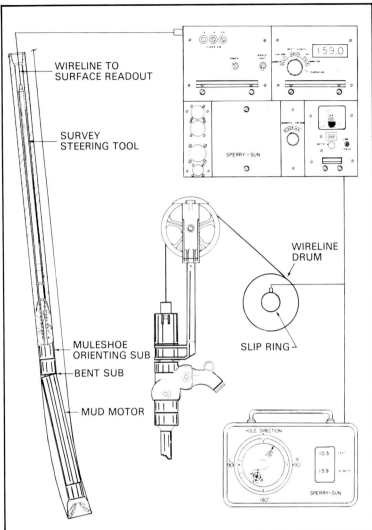

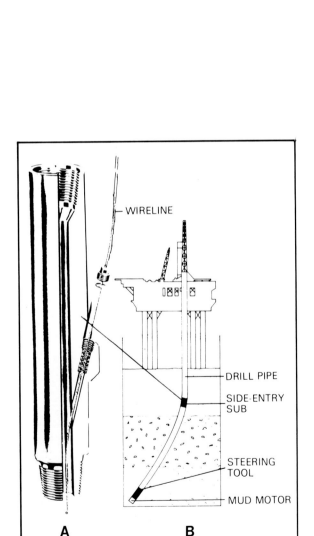

Figure 2.12. Side-entry sub: *A*, cutaway view; *B*, placement in drill string (*Courtesy of NL Sperry-Sun*)

A deflected downhole motor is often used in combination with a steering tool (fig. 2.11). The instrument is lowered into the nonmagnetic drill collar by wireline. At the surface, the wireline passes through a circulating head that, since the drill string is not rotated, is used in place of a kelly. The instrument must be tripped out to add pipe to the drill string. Usually, two or three joints are added at a time to avoid having to trip the instrument and reorient the tool so often.

Using a *side-entry sub* (fig. 2.12) allows the driller to add joints of drill pipe and deepen the hole without pulling the instrument. Near the surface, the wireline passes through the side-entry sub to the outside of the drill string, where it does not interfere with making up or breaking out joints of pipe. However, a split kelly bushing must be used to avoid fouling or cutting the wireline in the slips.

Sensors in the downhole instrument transmit data continuously, via the wireline, to the surface monitor. The operator can continuously read tool face orientation as well as hole azimuth and inclination. He can compensate for reactive torque, maintain hole direction, and change course when necessary without tripping out the drill string or interrupting drilling.

Orienting the Deflection Tool

To orient a deflection tool, the directional supervisor must know five things:
1. the present drift angle of the hole;
2. the present direction of the hole;
3. the drift angle desired at the end of the next section of hole;
4. the desired amount of change in direction; and
5. the rate at which the deflection tool can change hole angle.

With this information, he can determine the direction the tool should face, using either a digital computer or a type of slide rule called a *Ouija board*.

For example, suppose a bottomhole survey shows a drift angle and direction of 4° E. Suppose also that, according to the plan, the bottom of the next 100 ft of hole must have a drift angle of 5° in the direction S53°00'E. In other words, the hole angle must increase 1° and turn 37° to the right of its present course, toward the southeast instead of the east.

Since the hole is already slanting east, the directional supervisor must add a southerly component to the direction to bring it to the southeast. That is, he must run in a deflection tool and face it somewhere toward the south.

The Ouija board in figure 2.13 illustrates this problem. The new hole direction must be 37° different from the present direction; the new drift angle must be 5° instead of 4°. If the directional supervisor uses a downhole motor assembly that changes angle by 3° in 100 ft, he must face it 90° to the right of east – that is, due south.

Figure 2.13. Ouija board

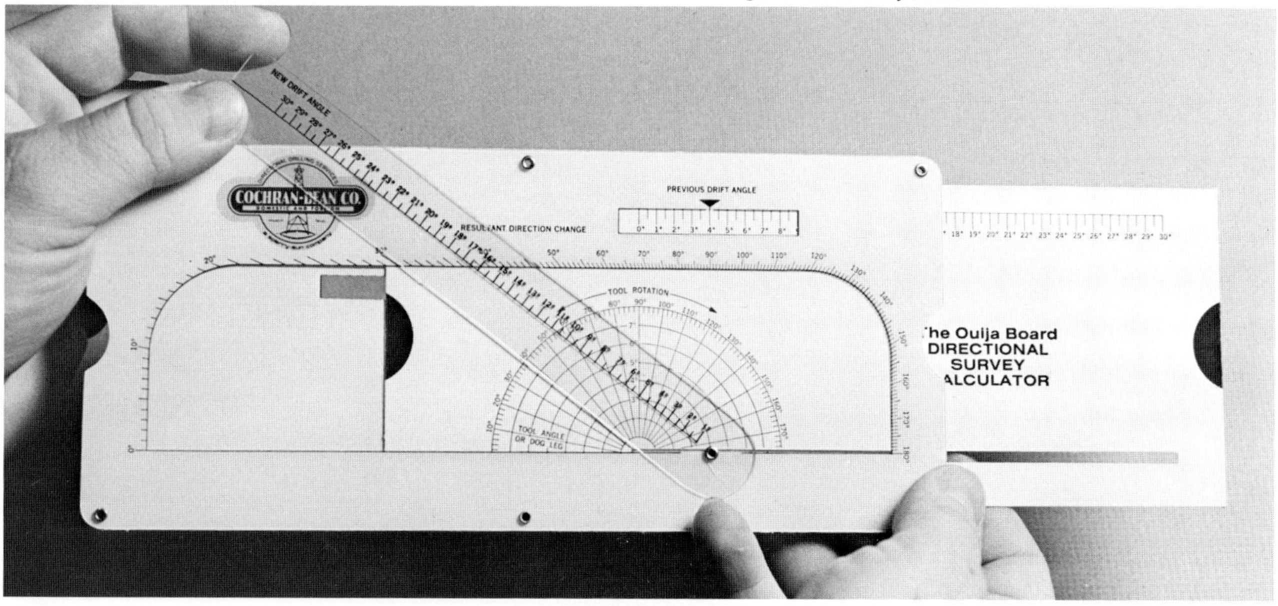

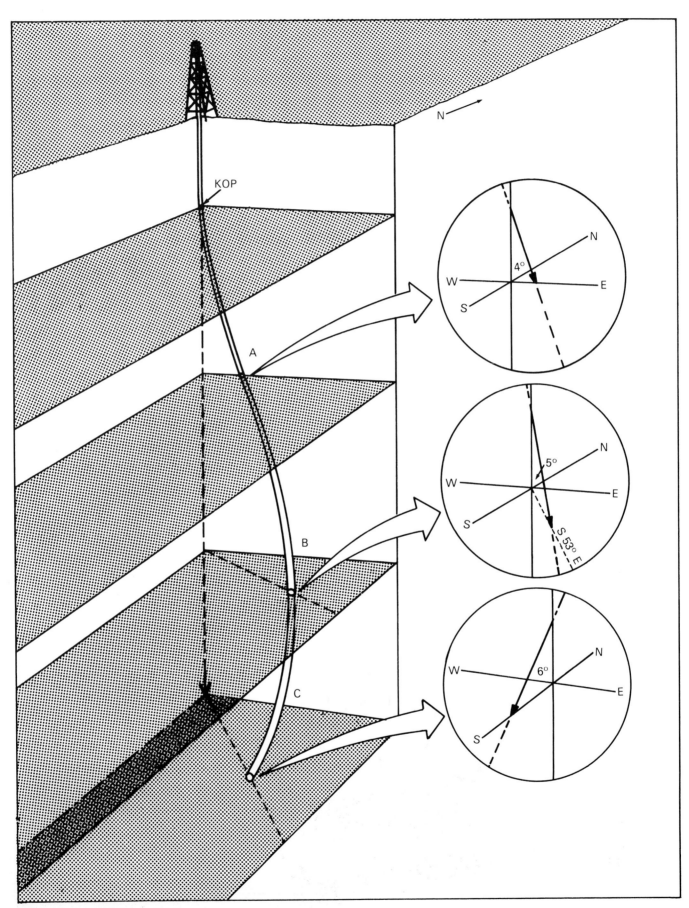

Figure 2.14. Wellbore deflection

Figure 2.14 shows what happens downhole. The first survey shows the direction and drift angle to be 4° due east (*A*). After beginning the deflection by facing the tool due south and drilling 100 ft of hole, the directional supervisor expects the next survey to show the hole drifting 5° S53°00'E (*B*).

Having determined the direction the tool must face, the directional supervisor orients the tool using either the *direct* or the *indirect* method. The *direct method* is used to orient tools in vertical or near-vertical holes or in holes whose inclination and direction are not known in advance. The *indirect method* can be used only where hole direction is known. Continuous surface readout using a rig-floor monitor is a modern and convenient form of direct orientation.

Direct method. When orienting the tool by the direct method, the directional supervisor determines the drift angle and direction of the hole at the same time that he determines the direction the tool is facing. A *muleshoe sub* (fig. 2.15) is made up as close as possible to the bit; when a downhole motor is used, it is placed between the bent sub and the first nonmagnetic collar. Inside the sub is a sleeve with a slot holding the *muleshoe key*. A photographic directional survey instrument is assembled with a muleshoe and a *stinger* on bottom. Inside the instrument is an *orientation line* that is aligned with the muleshoe keyway.

When the instrument assembly reaches the nonmagnetic collar, the stinger enters the muleshoe sleeve, causing the muleshoe on the instrument assembly to engage the muleshoe key projecting from the sleeve. The beveled edge of the muleshoe turns the instrument case so that it comes to rest with the key inside the muleshoe keyway.

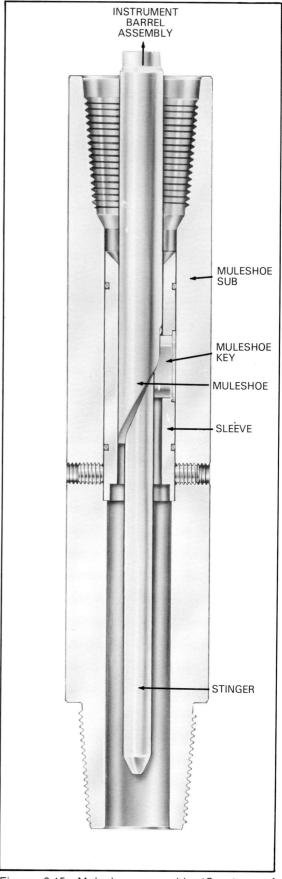

Figure 2.15. Muleshoe assembly (*Courtesy of Wilson Industries*)

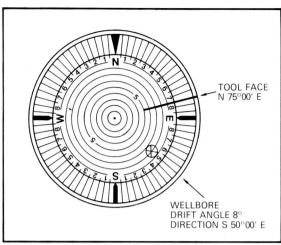

Figure 2.16. Single-shot magnetic orientation survey

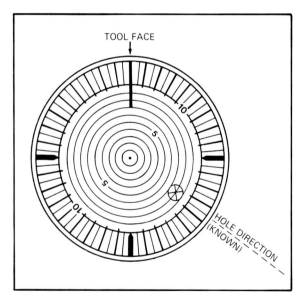

Figure 2.17. Orientation by indirect method

The orientation line is thus aligned automatically with the tool face. When the photographic disc is exposed, it records not only the drift angle and direction of the hole but also the direction the deflection tool is facing (fig. 2.16).

As in an ordinary directional survey, the magnetic compass readings must be corrected to true readings. However, a gyroscopic instrument, which provides true readings, can also be used for orientation.

If, as usually happens, the deflection tool has landed facing the wrong direction, it must be raised off bottom, turned, and set down again, and another orientation survey run. The process is repeated until the tool is properly aligned.

Steering tools save time by allowing the deflection tool to be oriented without interrupting drilling. The continuous surface readout shows not only the drift angle and azimuth of the hole but also tool face orientation. While drilling ahead with a downhole motor, the driller watches the surface readout and adjusts tool face orientation by turning the rotary table a few degrees left or right.

Indirect method. In indirect orientation, the deflection tool is oriented relative to the low side of the hole, which is determined by a directional survey before the tool is run into the hole. The hole must be inclined at least 1° to 3° (depending upon the instrument used) for accurate orientation. As in indirect orientation, the instrument is aligned with the tool face by a muleshoe assembly.

When the tool comes to rest on the bottom and the photographic disc is exposed and developed, it shows the direction the tool is facing with respect to hole direction (fig. 2.17). In the example, the tool has landed facing 125° counterclockwise from the hole direction. As in direct orientation, the process is repeated until the tool lands facing the desired direction.

Bottomhole Assemblies

It is common practice now to drill long sections of curved or straight hole with downhole motors. However, for economic or other reasons, it may be advantageous to drill as much of a deviated well as possible with a conventional rotating drill string. If the hole is not curving at a satisfactory rate, the drill string may have to be tripped out and modified. Making a special trip to run in a deflection tool can often be avoided by using an appropriate bottomhole assembly and adjusting weight on bit, rotary speed, and circulation.

The term *bottomhole assembly* (*BHA*) refers to a combination of drill collars, stabilizers, and associated equipment made up just above the bit. In directional drilling, particularly during rotating, the BHA used affects whether hole angle increases, decreases, or remains the same. Except with a jet bit, a rotating BHA cannot be used to control horizontal direction or to kick off a directional hole; but specific BHAs are useful for changing the drift angle of a wellbore that is already deviated.

All parts of the drill string are flexible to some degree. Standard drill pipe is very limber and bends easily under compression; for that reason the upper drill string is normally kept in tension while drilling. But even large, heavy drill collars used at the bottom of the drill string are limber enough to bend where they are unsupported. Ideally, altering the BHA in a particular way enables the driller to control the amount and direction of bending and thereby to increase, decrease, or maintain drift angle as desired.

Drilling large-diameter directional holes – 8½ to 12¼ inches – is usually easier than drilling small-diameter holes. Larger drill collars and pipe are stiffer and therefore less subject to sagging or bending and being thrown off course by the formations being drilled. They also weigh more, giving the driller a greater usable weight-on-bit range. Although their large surface area makes them more likely to become wall-stuck, their advantages so outweigh their disadvantages that their use has become standard practice in directional drilling.

Fulcrum assembly. A stabilizer made up just above the bit acts as a *fulcrum*. In holes inclined 3° or more off vertical, the drill collars above the fulcrum sag toward the low side, forcing the bit toward the high side and increasing hole angle as drilling progresses (fig. 2.18). This tendency is called the *fulcrum effect*. BHAs using the fulcrum effect to increase hole angle are also called *building assemblies*.

The rate of angle buildup with a fulcrum assembly can be controlled by selecting an appropriate collar size, using short collars or subs between the stabilizer and the bit, distributing other stabilizers appropriately above the bottom drill collars, and adjusting weight on bit and circulation rate. The more limber the assembly is above the fulcrum, the faster the hole angle builds. Smaller-diameter drill collars sag and bend more readily than larger-diameter collars. Applying more weight causes the BHA to bend further in the direction of the initial sag. Using only enough circulation to clean the bit and stabilizer builds angle fastest, especially in soft formations.

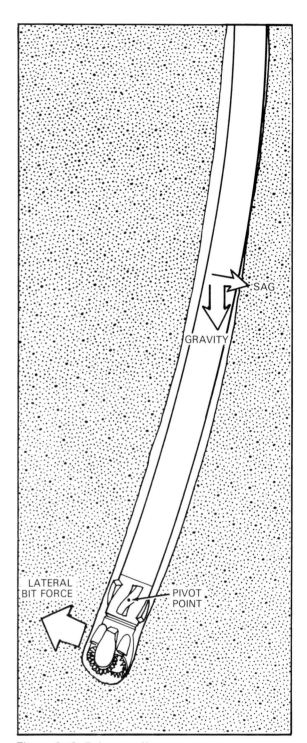

Figure 2.18. Fulcrum effect

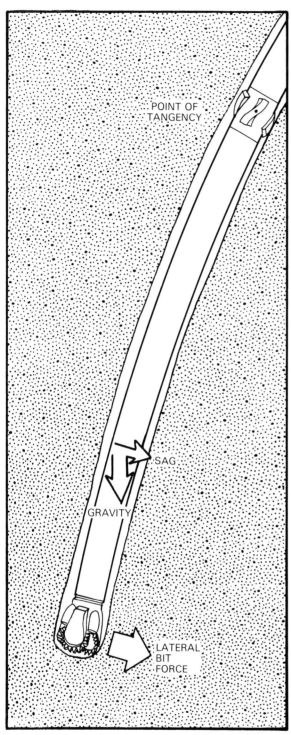

Figure 2.19. Pendulum effect

Pendulum assembly. In a limber assembly supported by a stabilizer one or two collars above the bit instead of a near-bit stabilizer, the drill collars tend to hang vertically below the stabilizer (fig. 2.19). In a slanted hole, gravity forces the bit against the low side, tending to decrease hole angle as drilling progresses. This principle is known as the *pendulum effect*. A pendulum BHA used to reduce drift angle is sometimes called a *drop assembly*. (The pendulum effect is also used to maintain a straight course in crooked-hole country.)

In a pendulum assembly, the distance that the stabilizer must be placed above the bit depends upon the stiffness of the drill collars. If a smaller-diameter, limber collar is used, the stabilizer should be placed lower in the drill string to keep the collar from sagging against the hole wall. The pendulum effect is nullified if the drill collar contacts the low side between the bit and the stabilizer. Smaller-diameter collars also exert less weight on the bit, reducing the rate of penetration. To prevent sagging in highly deviated holes, the stabilizer may have to be placed so low that the bit exerts little or no force on the low side. In some cases, an undergauge stabilizer placed near the bit will cause a gradual drop in angle; however, if all else fails and angle cannot be decreased as desired, it may become necessary to trip out and run in a deflection tool.

To continue making hole at a satisfactory rate, a driller may have to use faster rotary speeds; but faster rotation may also increase the tendency of the bit to deviate the hole, depending on hole inclination, geologic structures being penetrated, and other factors. To drop angle at a satisfactory rate, the driller must maintain a delicate balance between the desired rates of penetration and dropoff.

Packed-hole assembly. Doubling the cross-sectional area of a collar increases its stiffness eight times. To maintain hole angle, the driller may use a combination of large, heavy drill collars and stabilizers to minimize or eliminate bending, thus eliminating both the pendulum and the fulcrum effects. Such a BHA is called a *packed-hole assembly* or *stiff assembly* (fig. 2.20).

Downhole motor assembly. A downhole motor may be used not only to change hole angle and direction but also to drill the straight (vertical or inclined) sections of a directional hole. When a downhole motor is used to maintain angle, blade-type stabilizer ribs may be welded onto the lower end of the housing and a full stabilizer made up just above it. To eliminate static friction and transmit the specified weight to the bit, the drill string is sometimes rotated slowly when a downhole motor assembly is being used to drill straight ahead—that is, when a bent sub or bent motor housing is not being used.

SPECIAL PROBLEMS IN DIRECTIONAL DRILLING

Directional wells are harder to drill than straight holes. Nearly everything done in routine drilling becomes more complicated when the well has to be drilled at an angle. Raising and lowering the drill string requires more hoisting capacity; greater rotary torque is needed to overcome friction; mud and hydraulic system requirements are more critical; stuck pipe and equipment failures are more common; and casing is harder to run and cement.

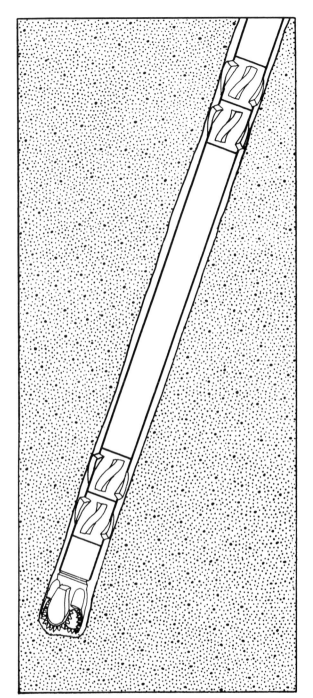

Figure 2.20. Stiff bottomhole assembly

Dogleg Severity

Many problems can be avoided by paying special attention to the rate of angle change. Ideally, drift angle should be built or dropped gradually: 2°/100 ft is standard, with a safe maximum of about 5°/100 ft. However, changing drift angle from 3° to 7° or 8° is not automatically safe. Angle change must be evenly distributed over the course length. If 3° of angle is added smoothly over 100 ft and the horizontal direction does not change, probably no trouble will result during subsequent drilling or production. However, if the buildup occurs in the first 50 ft, with the last 50 ft staying straight, the rate of build in the first 50 ft is

$$3° \times \frac{100}{50} = 6°/100 \text{ ft.}$$

The *dogleg severity factor* is more complex. Both vertical and horizontal changes must be considered, along with average inclination (fig. 2.21). If inclination builds smoothly from 8° to 12° over 80 ft, the buildup rate is 5°/100 ft. But if hole direction changes 25° at the same

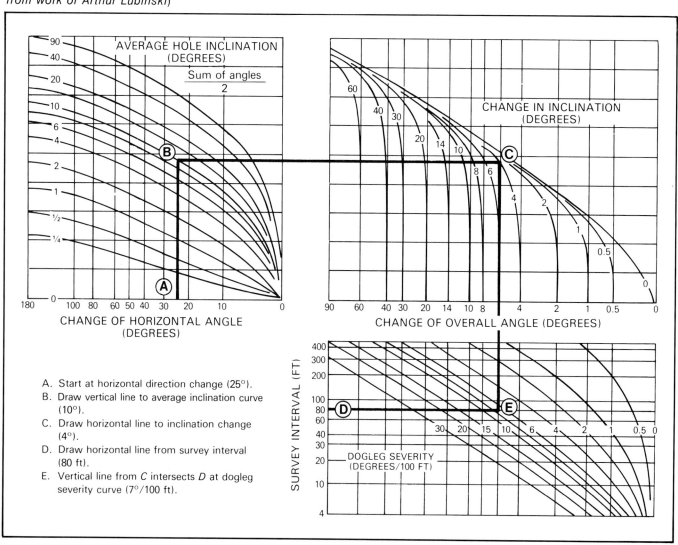

Figure 2.21. Determining dogleg severity (*Adapted from work of Arthur Lubinski*)

A. Start at horizontal direction change (25°).
B. Draw vertical line to average inclination curve (10°).
C. Draw horizontal line to inclination change (4°).
D. Draw horizontal line from survey interval (80 ft).
E. Vertical line from C intersects D at dogleg severity curve (7°/100 ft).

time, the dogleg severity factor becomes 7°/100 ft and the hole has a spiral or corkscrew shape.

Severe doglegs in the upper part of the hole may cause *key seating* (fig. 2.22). The weight of the drill string below the dogleg forces pipe against the side of the hole, wearing an undergauge groove too small for a tool joint or drill collar to pass through. When the drill string is raised or lowered, it may become stuck in the key seat and have to be fished out, an expensive and time-consuming operation. If the hole is cased, the casing may be worn through while the lower part of the hole is being drilled. For these reasons, it is safer to build angle rapidly in the lower part of the hole than in the upper part.

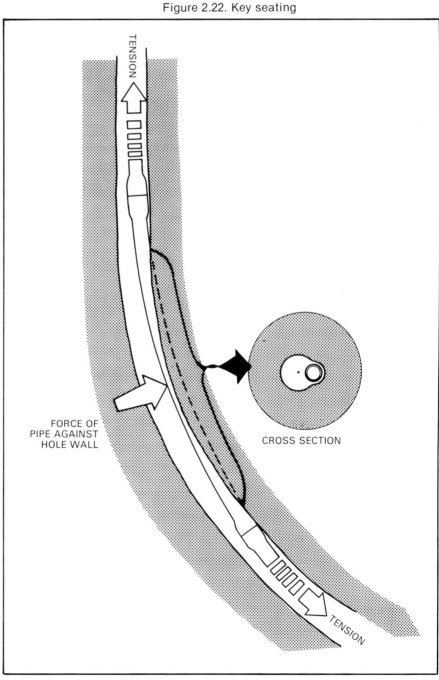

Figure 2.22. Key seating

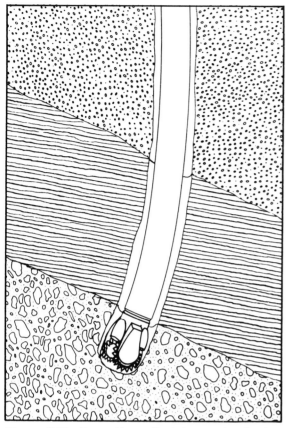

Figure 2.23. Formation effect in strata dipping less than 45°

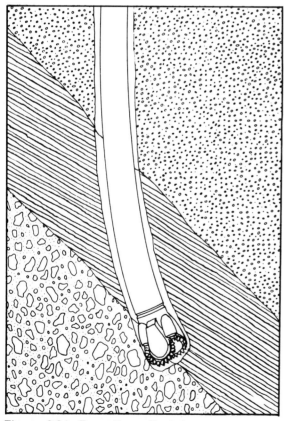

Figure 2.24. Formation effect in strata dipping more than 45°

Formation Effect

Sometimes the formation tends to deflect the bit. Directional control becomes more difficult in drilling through laminar formations that are not level. When the dip (the formation's angle off the horizontal) is less than about 45°, the bit tends to drill updip, or perpendicular to the layers (fig. 2.23); when dip is more than 45°, the bit drills downdip, or parallel to the layers (fig. 2.24). Sometimes a well can be planned to exploit or compensate for this tendency. Otherwise, a stiff BHA must be used to overcome formation effect.

The drill bit also tends to deviate horizontally parallel to tilted strata. This effect is called *wandering*. Even where strata are horizontal, the right-rotating bit tends to walk to the right in inclined holes (*bit walk*). Wandering and bit walk are harder to control than vertical directional changes because they cannot be corrected simply by changing the rotary BHA.

If a stiff assembly fails to control bit walk or wandering, a deflection tool is usually required; a steering tool can be a great time- and money-saver in this situation. In most instances, however, the driller can anticipate formation effect or bit walk and compensate for it by leading the hole—that is, kicking off in a direction other than that shown on the plan (usually to the left) and letting the bit walk into the target.

Hydraulics

As a rule, directional drilling is most efficient with a high penetration rate, which requires high circulation pressures to clean out cuttings. However, to achieve the best overall results at the lowest possible cost, the driller must balance many factors, such as penetration rate, angle buildup, the likelihood of drill stem abrasion or sticking, and pressure control. Penetration rate, for instance, is limited when building angle because both weight on bit and circulating pressure must be restricted to control the rate of deflection.

In the inclined section, especially in high-angle holes, cuttings tend to settle to the low side (fig. 2.25). The drill string also sags to the low side, making the problem worse by interfering with return circulation. Stabilizers help alleviate the problem by holding the drill string away from the hole wall; otherwise, the only way to compensate is to increase the carrying capacity of the mud. This can be done by adjusting the viscosity or density of the drilling fluid or increasing the circulation rate. Formation pressure control or other considerations may not allow increases in mud weight or viscosity; on the other hand, increasing circulation rate may cause a building assembly to deviate too quickly or to wander.

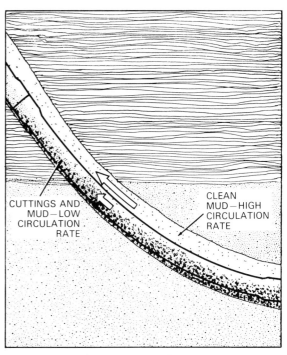

Figure 2.25. Circulation problem in high-angle holes

Friction

In high-angle holes, much or most of the drill string weight is directed against the low side of the hole (fig. 2.26). The resulting friction requires more power to rotate the drill string and increases tool joint and pipe wear and failure. In soft formations, it may even cause key seating on the low side. Drill string friction cannot be eliminated, but it may be reduced by circulating with oil-emulsion muds.

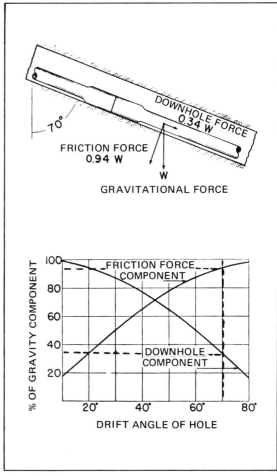

Figure 2.26. Friction problem in high-angle holes

Figure 2.27. Casing centralizers

Friction also makes it difficult to run casing down directional holes. Casing lying on the low side also results in poor cement distribution. Using flush-joint casing and centralizers reduces wall friction; centralizers also allow proper cement distribution by holding casing away from the wall (fig. 2.27).

In high-angle holes, friction may keep directional surveying instruments from sliding freely down the drill string. The driller may have to start the mud pump and use circulating pressure to force the instrument to the bit.

SUMMARY

In the real world, the directional supervisor must consider many factors not discussed in this lesson, such as formation stability, blowout prevention, drill string wear or failure, equipment malfunctions, and inexplicable random errors and effects. He works blind, trying to guide a spinning, wobbling string of metal a few inches thick and thousands of feet long into an area perhaps 200 ft wide and 2 miles away. It is not unlike trying to thread a strand of cooked spaghetti into the depths of an anthill in search of a particular ant. That it is possible at all testifies to the state of drilling technology and the directional driller's art.

LESSON 2 QUESTIONS

Put the letter for the best answer in the blank before each question.

Questions 1–21 cover material discussed in **Changing the Course of the Hole.**

_____ 1. The most important factor in choosing a deflection tool is –
   A. the amount of deflection needed.
   B. the formation to be drilled.
   C. the depth of the hole.
   D. the temperature at the bottom of the hole.

_____ 2. *Tool face* is –
   A. the bottom of the drill bit.
   B. the top of the uppermost drill collar.
   C. the side of the deflection tool toward which the bit drills.
   D. magnetic north.

_____ 3. The tool face is always oriented in the desired hole direction. (T/F)

_____ 4. Whipstocks can be used –
   A. to bypass a stuck fish.
   B. to sidetrack out of cased hole.
   C. in open hole.
   D. in any of the above situations.

_____ 5. A whipstock can be used to drill 200 feet of hole in a single operation. (T/F)

_____ 6. A jet bit can be used to change hole direction and then, without tripping, rotated to drill ahead. (T/F)

_____ 7. Which of the following statements about downhole motors is true?
   A. They drill only undergauge hole.
   B. They drill without circulation.
   C. They must be tripped out to adjust hole angle.
   D. They drill at relatively high rotational speeds.

_____ 8. Positive-displacement downhole motors can be powered by –
   A. air.
   B. drilling mud.
   C. water.
   D. any of the above fluids.

_____ 9. Reactive torque must be compensated for during use of—
   A. a positive-displacement motor.
   B. a whipstock.
   C. a jet bit.
   D. mud-pulse telemetry.

_____ 10. According to the rule of thumb used in compensating for reactive torque (p. 43), when drilling at 3,500 ft MD in hard rock, the driller should orient a downhole motor _____ to the right of the desired direction.
   A. 3½°
   B. 5°
   C. 10°
   D. 17½°

_____ 11. A bottomhole directional reading of 15° S10°30'W means that—
   A. the bottom of the hole is slanting 10½ degrees.
   B. the bottom of the hole is located 15 degrees away from a vertical line beneath the rig.
   C. the bottom of the hole is slanting 15 degrees.
   D. the bottom of the hole is located 30 feet southwest of the rig.

_____ 12. To orient a deflection tool by the indirect method, the driller must have prior knowledge of bottomhole drift angle and direction. (T/F)

The diagram at right shows the result of an orientation survey by the direct method, using muleshoe alignment. Read the diagram and answer questions 13–15 by placing the letter of the correct answer on the right in the appropriate blank.

_____ 13. Drift angle   A. S70°W
_____ 14. Hole direction   B. S70°E
                             C. N30°W
_____ 15. Tool face direction D. 7°
                             E. 8°
                             F. 30°

_____ 16. Directional control is easier in small-diameter holes than in large-diameter holes. (T/F)

_____ 17. A pendulum BHA is used to—
   A. change hole direction to the left.
   B. change hole direction to the right.
   C. increase drift angle.
   D. decrease drift angle.

_____ 18. An angle-building assembly usually includes—
   A. a stabilizer immediately above the bit.
   B. a long, limber drill collar between the bit and the first stabilizer.
   C. no stabilizers at all.
   D. no drill collars.

_____ 19. A packed-hole assembly—
   A. is used to increase drift angle.
   B. is used to drop angle rapidly.
   C. is used to maintain drift angle.
   D. contains no stabilizers.

_____ 20. If drill collar *A* has twice the cross-sectional area of drill collar *B*, then *B* is _____ as stiff as *A*.
   A. ⅛
   B. half
   C. twice
   D. 8 times

_____ 21. The least flexible type of bottomhole assembly is the—
   A. fulcrum assembly.
   B. packed-hole assembly.
   C. muleshoe assembly.
   D. pendulum assembly.

If you need to, review **Special Problems in Directional Drilling** before answering questions 22–27.

_____ 22. If drift angle decreases from 24° to 19° and direction changes from S45°W to S75°E in 100 ft of hole, the dogleg severity factor is 5°/100 ft. (T/F)

_____ 23. Key seating is most likely to occur—
   A. in the lower part of the hole.
   B. in the upper part of the hole.
   C. in straight hole.
   D. in cased hole.

_____ 24. When drilling through rock layers slanted less than 45°, the bit tends to—
   A. drill vertically.
   B. drill parallel to the layers.
   C. drill perpendicular to the layers.
   D. drill out a key seat.

_____ 25. Hole angle and direction are 25° N45°00'E in a horizontally stratified formation. In which direction would you expect the bit to walk?
   A. Updip
   B. Counterclockwise
   C. Right
   D. Vertically

_____ 26. Methods of improving hole cleaning in high-angle holes include—
  A. using stabilizers.
  B. adjusting mud density.
  C. increasing circulation rate.
  D. all of the above methods.

_____ 27. Using centralizers on casing—
  A. improves cement distribution.
  B. prevents key seating.
  C. reduces drill string friction.
  D. helps in running survey instruments down to the bit.

As a review exercise for this lesson, match the terms on the right with the descriptions on the left by placing the letters in the appropriate blanks. Use each letter only once, and only one letter per blank.

_____ 28. Joint with pin and box on different axes

_____ 29. Tendency of bit to deviate to right

_____ 30. Type of BHA used to increase drift angle

_____ 31. Horizontal formation effect

_____ 32. Deflection tool that washes out side of hole

_____ 33. Nonrotating part of downhole motor

_____ 34. Bottomhole assembly

_____ 35. BHA used to maintain hole angle

_____ 36. Side of deflection tool to which bit drills

_____ 37. Tendency of drill collars to hang vertically

_____ 38. Rotation of mud motor opposite bit rotation

_____ 39. Wedge-shaped deflection tool

_____ 40. Undergauge slot in hole wall

A. Bit walk
B. Pendulum effect
C. Whipstock
D. BHA
E. Jet deflection bit
F. Fulcrum
G. Key seat
H. Wandering
I. Tool face
J. Reactive torque
K. Bent sub
L. Stiff assembly
M. Stator

Lesson 3
Open-Hole Fishing

Introduction

Causes of Fishing Jobs

Fishing Equipment and Techniques

The Economics of Fishing

Lesson 3
OPEN-HOLE FISHING

INTRODUCTION

Losing equipment in the hole is one of the most expensive things that can go wrong in drilling a well. Drilling must come to a halt while the lost equipment is recovered, or the hole must be sidetracked around it. Control of the well may become hard to maintain with essential tools out of reach, and a blowout is both expensive and dangerous.

The term *fishing* was probably coined by a cable-tool driller using a barbed spear to snag a broken cable in the depths of a narrow wellbore. The retrieved cable-tool bit or bailer dangling on the end of its line inevitably became known as a *fish*. But in these days of rotary drilling, a fish can be any undesirable object in the hole that must be removed by a special operation before drilling can continue, such as a lost wireline logging tool, a bit cone, hand tools, a large part of the drill string, or even naturally occurring pieces of iron pyrite.

fish

To remove a fish from the hole, specialized equipment and procedures must be used. The job may also require specialized knowledge and experience. Improper use of tools or procedures may turn a simple fishing job into an expensive project. Each fishing job is unique. The tools and techniques used to fish a string of stuck drill pipe from a well may not work at another well or under other conditions at the same well. However, this lesson will describe some of the basic techniques and tools used in open-hole fishing – that is, retrieving fish from a hole that is being drilled but is not yet cased.

CAUSES OF FISHING JOBS

Just as there are many different types of fish, there are many different ways objects can become lost or stuck in the hole. The biggest type of fish is a portion of the drill string that has become stuck and either has broken off or has been purposely disconnected to recover the pipe above it.

Relatively small fish, known as *junk*, can also result from drill string failure. Slivers of metal may come loose when the drill string parts. Metal fragments produced by grinding, or *milling*, a fish to aid in its recovery may impede drilling after the larger item has been fished out; metallic fragments are especially hard on diamond bits. Junk from uphole may also stick the drill string by jamming between the drill pipe or collars and the hole wall.

junk

milling

The most common causes of fishing jobs, not necessarily in order of frequency, can be classified as follows:
1. *twistoff*, or parting of the drill string caused by metal fatigue;
2. sticking of the drill string, sometimes resulting in a twistoff;
3. bit and tool failures; and
4. foreign objects, such as hand tools, logging instruments, broken wireline or cable.

Twistoff

When drill pipe parts during normal drilling operations, the cause is usually not excessive torque but metal fatigue. Rough handling, scarring by tong dies, improper makeup torque, and other types of damage create weak spots where cracks can form and enlarge under the constant bending and torquing stresses of routine drilling. The pipe often separates in a characteristic helical break (fig. 3.1) or in a long tear or split.

Surface signs of a twistoff include any or all of the following:
1. loss of drill string weight;
2. lack of penetration;
3. reduced pump pressure;
4. increased pump speed;
5. reduced drilling torque; and
6. increased rotary speed.

Stuck Pipe

A drill string thousands of feet long can get stuck in many places and many ways in a hole only a foot or so across. There are two general categories of drill string sticking—*mechanical* and *differential*. In mechanical sticking, the drill string is lodged in place by solid material; in differential sticking, by fluid pressure.

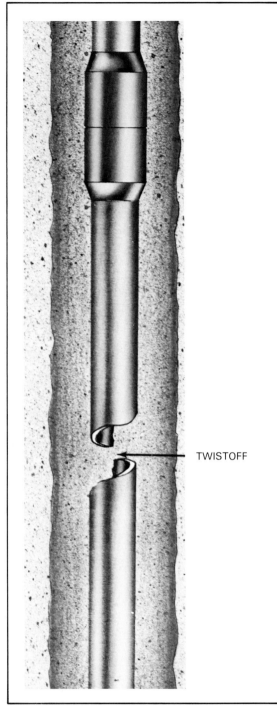

Figure 3.1. Parted pipe showing fatigue break

Mechanical sticking. Among the ways the drill string can become mechanically stuck are the following:

1. *Sloughing hole* (fig. 3.2) is often the result of shale absorbing water from the drilling fluid, expanding, sloughing off, and falling downhole. Large masses may lodge around drill collars and stabilizers, sticking the drill string and blocking circulation. Abnormally pressured shale, steeply dipping shale beds, and erosion by drilling fluid can also cause the hole wall to cave in.

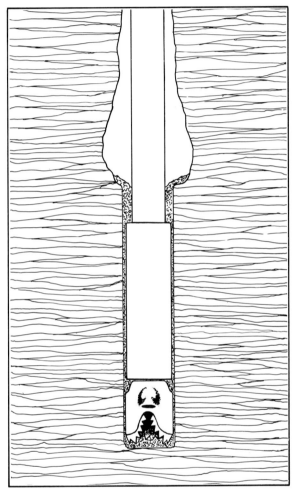

Figure 3.2. Pipe stuck in sloughing hole

2. *Undergauge hole* (fig. 3.3) is often associated with shale also. If the formation swells but does not slough off, the deformed layer may close around the drill pipe, cutting off circulation and preventing passage of the larger-diameter tool joints, drill collars, or bit. A buildup of mud solids can have the same effect, especially in a permeable zone where water is lost to the formation.

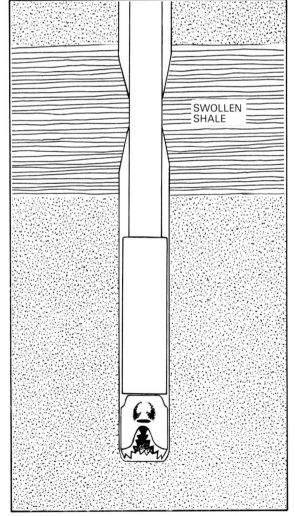

Figure 3.3. Pipe stuck in undergauge hole

3. *Blowout sticking* (fig. 3.4) occurs when a large volume of sand or shale is driven uphole by formation fluids entering the wellbore.

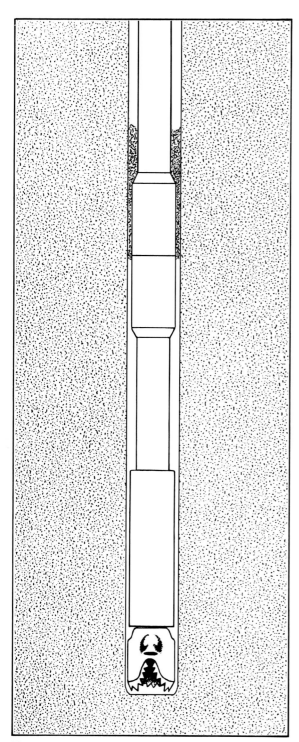

Figure 3.4. Pipe stuck by blowout

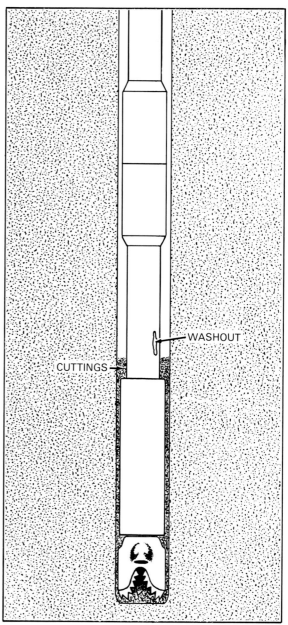

Figure 3.5 Pipe stuck because of drill string washout

4. *Inadequate hole cleaning*, failure of the circulating system to clean cuttings or other material from the hole, can be the result of sloughing shale, drill string washout above the bit (fig. 3.5), a low circulation rate in a large hole with unweighted mud, or lost returns. The result is a buildup of solids around the bit and collars.

5. *Junk in the hole* (fig. 3.6), such as broken-off or dropped equipment, may lodge between hole wall and drill pipe, tool joints, or drill collars.

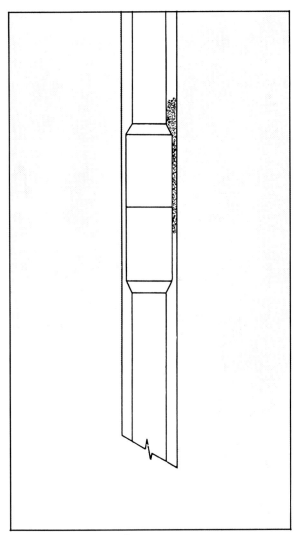

Figure 3.6. Pipe stuck by junk in hole

6. *Key seating* (fig. 3.7) occurs when drill pipe in tension wears an undergauge groove in the wall of a curved section or dogleg. When the drill string is raised or lowered, tool joints or drill collars may become lodged in the lower or upper end of the key seat.

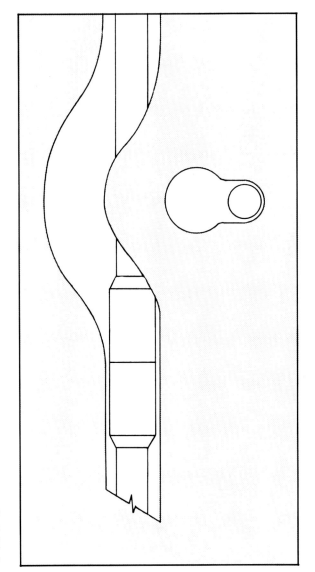

Figure 3.7. Pipe stuck in key seat

67

7. *Tapered hole* is the result of bit gauge wear in drilling hard, abrasive formation. Tripping in a new bit without reaming it to bottom can jam it partway down the tapered section (fig. 3.8).

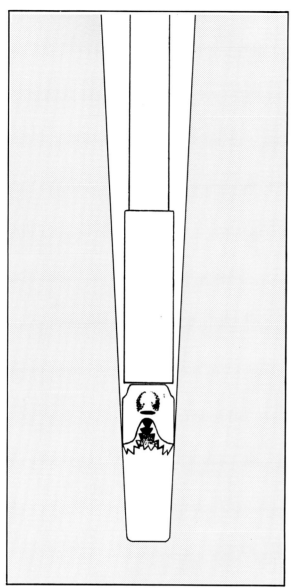

Figure 3.8. Bit stuck in tapered hole

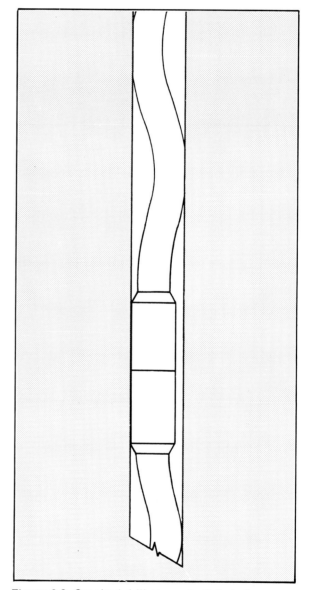

Figure 3.9. Crooked drill stem stuck in hole

8. *Crooked pipe* (fig. 3.9), often the result of dropping the drill string or applying excessive weight to stuck pipe, may jam against the walls of the hole and become impossible to raise, lower, or rotate.

Differential sticking. Many fishing jobs are caused by drill pipe or collars becoming differentially stuck, or *wall-stuck*. Wall sticking commonly occurs in permeable formation where hydrostatic pressure of the drilling mud forces mud liquids into the lower-pressure formation and causes a buildup of mud solids known as *wall cake* or *filter cake*. The pressure of mud in the annulus presses the pipe against the wall cake lining the hole (fig. 3.10). This pressure may be strong enough to support the entire drill string. The longer the drill string remains motionless, the more likely it is to become wall stuck and the harder it is to free. Excessive hydrostatic pressure may be responsible, but only a significant overbalance, or pressure differential from wellbore to formation, is necessary for wall sticking to occur. The more solids that are in the mud, the thicker will be the wall cake – and the tighter the pipe will be stuck.

If the pipe becomes stuck while motionless in a clean, well-conditioned hole and circulation can be stopped and then resumed at the original pressure, the chances are that the pipe is wall-stuck. Prompt action, such as immediately rotating the drill string, may free wall-stuck pipe that cannot be pulled free. *Oil spotting* (pumping a slug of oil-based

wall sticking

wall cake
filter cake

oil spotting

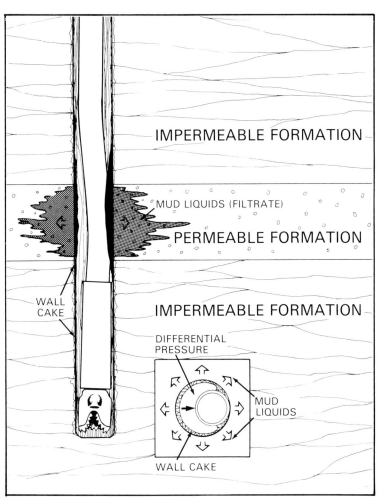

Figure 3.10. Differential sticking

fluid down the drill stem and up the annulus to the stuck point) may break the mud seal. If these actions do not free the pipe, a fishing job may become necessary. Using heavy-weight drill pipe, each joint of which has a large external upset in its center, helps prevent wall sticking.

Other Fishing Jobs

Smaller fish, or junk, may include such items as—
1. bit cones, bearings, or other parts lost when a bit breaks;
2. broken reamer or stabilizer parts;
3. fragments of metal lost in a twistoff;
4. metal fragments produced by milling the top of a fish to aid in its retrieval;
5. naturally occurring pieces of hard, crystalline, or abrasive minerals, such as iron pyrite;
6. tong pins, wrenches, or other items that fall into the hole due to rig equipment failure, accident, or carelessness; and
7. equipment such as packers, corers, and drill stem test (DST) tools that have become lodged downhole.

In addition, wireline equipment can become lost or stuck downhole. If the wireline is intact, the fishing job is to dislodge the tool; if the wireline has parted, the main fishing job may be to snag the wireline.

FISHING EQUIPMENT AND TECHNIQUES

When it becomes necessary to fish drilling equipment out of a hole, the experienced operator finds out as much as possible about the situation before taking action. Among the questions he may attempt to answer are the following:
1. What is to be fished out of the hole?
2. Is the fish stuck, or is it resting freely?
3. Is it in casing or in open hole?
4. If stuck, what is causing it to stick?
5. What is the condition of the hole?
6. What is the condition of the fish?
7. Could fishing tools be run inside the fish, or must they be run outside it?
8. Could other tools be run through the fishing assembly that is to be used?
9. Are there at least two ways to get loose from the fish if it cannot be freed?

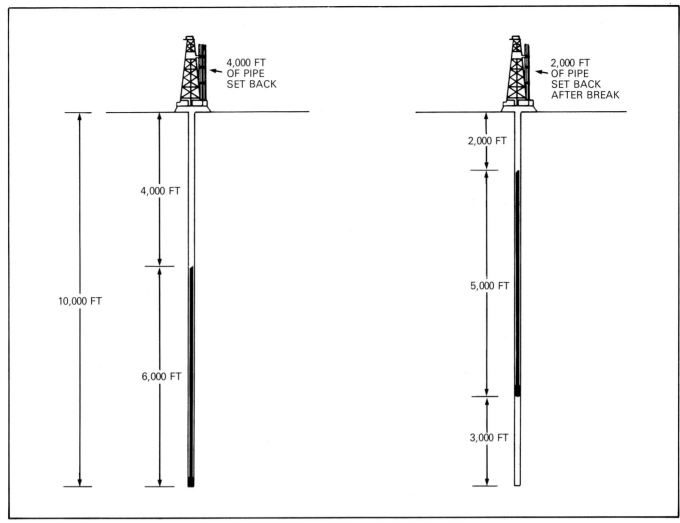

Figure 3.11. Finding depth of top of fish by measuring drill string removed from hole

In a fishing job involving the drill string, the operator can often ascertain whether or not the lost drill pipe is stuck in the hole by determining what happened just before it was lost. For instance, if the bit was on bottom and drilling, and if there was no sudden, unexplained increase of torque or decrease in rotary speed before the drill string broke, the most likely explanation is the occurrence of a twistoff, and the pipe is probably not stuck. On the other hand, if the pipe was motionless in the hole or if it was being raised or lowered but not rotated, it is probably stuck – either mechanically or differentially.

The operator must know, as accurately as possible, the depth at which the top of a broken-off drill string can be found. The upper section of the string is measured as it is removed from the hole. If the bit was on bottom when the drill string broke, or if the drill string became stuck off bottom, the length of the upper portion is the same as the measured depth of the top of the fish (fig. 3.11).

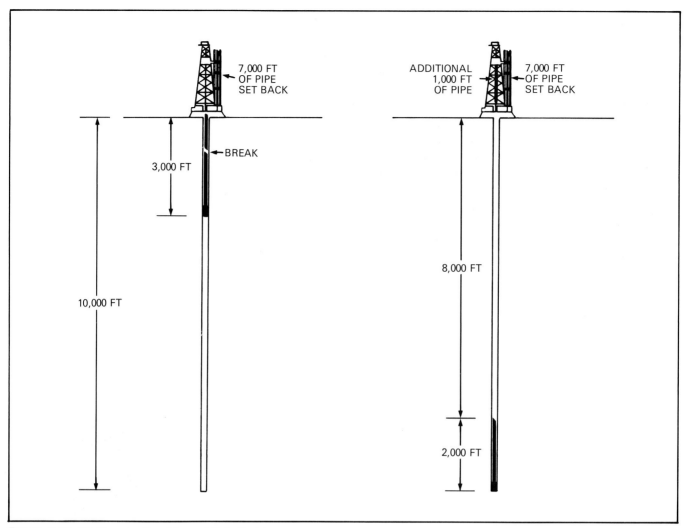

Figure 3.12. Finding depth of top of fish that has fallen to bottom

On the other hand, if the drill string broke with the bit off bottom and the fish then fell downhole, the remainder of the drill string in the hole must be measured as it is set back. In this way, the depth of the top of the fish can be closely estimated, assuming that the fish fell all the way to the bottom (fig. 3.12).

By determining the depth of the top of the fish, the operator can ascertain whether the fish is in open or cased hole. If there is any doubt about the fish's location, the operator can run an electric log.

Hole condition is an important consideration in recovering a fish. If drill pipe is stuck by a cave-in or by swelling shale, circulation may be restricted or cut off altogether. On the other hand, circulation is usually not affected when the drill string is stuck in a key seat or when pipe is wall-stuck. Small fish may be buried by cuttings or cavings. Hole condition is different in each situation, but circulation is useful, and often necessary, in correcting the problem.

The condition of the fish is essential information. Most pipe recovery tools are designed with close tolerances; specific component sizes are needed for specific jobs. Irregularities likely to hamper recovery of a fish must be dealt with first. If the top of a section of broken-off drill pipe is burred or splayed, it may be necessary to clean up or *dress* the pipe before trying to pick it up. Junk may have to be broken apart in order to be recovered.

Inspecting the upper portion of the break as it is pulled from the hole may provide a good reverse image of the top of a twistoff. Another way to get an idea of the condition of the top of a fish is to run an *impression block*. A typical impression block consists of a block of lead, with a circulation port, molded onto a steel body (fig. 3.13). The block is made up on drill pipe and collars and run into the hole until it is just above the fish. Circulation is started to wash all settlings off the top of the fish so that a good impression can be obtained. The block is lowered gently to contact the fish; then weight is applied. The top of the fish indents the bottom of the soft lead block, leaving an impression that can be examined at the surface.

Figure 3.13. Impression block

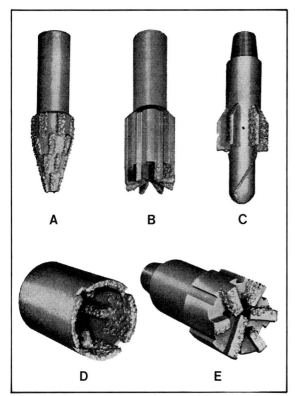

Figure 3.14. Types of mills: *A*, tapered; *B*, concave; *C*, piloted; *D*, cleanout; *E*, flat-bottomed

Fishing Out a Twistoff

If part of the drill string has broken off in open hole and it is not stuck, the fishing job consists mainly of locating and engaging the top of the fish with an appropriate fishing tool.

Suppose the operator finds the top of the broken-off pipe badly twisted. The tools he is going to use need a section of straight, undamaged pipe to make a firm, positive catch. He will have to remove the damaged metal to give the fish a more acceptable shape, preferably a cross section such as an upset or a tool joint.

Depending upon the fish's shape, any of a variety of *mills* can be used. Mills (fig. 3.14) have cutting surfaces inlaid with tungsten carbide, a hard, abrasive alloy that grinds away even high-strength steel. Mills can be used to dress the top of a fish or to grind away entirely a stuck fish that cannot be retrieved by any other method.

The drill string must be rotated slowly and carefully. It may be necessary to use a *piloted mill* (fig. 3.15), which will not jump off the top of a fish and go down beside it. High-volume circulation should be maintained to flush the cuttings and cool the mill.

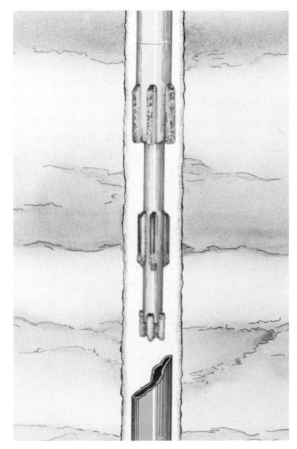

Figure 3.15. Piloted mill

Once he has determined that the top of the fish is fairly smooth, the operator makes up a *fishing string*. A typical fishing string (fig. 3.16) consists of an overshot, a jar, a series of drill collars, and a jar accelerator made up on drill pipe. To calculate how much pipe and how many collars he needs to reach the fish, he must consider the made-up length of the fishing tool assembly, as well as the distance the fish must travel inside the assembly before it can be firmly caught.

A typical *circulating and releasing overshot* (fig. 3.17) consists of three outside parts: the top sub, the bowl, and the guide. The top sub connects the overshot to the fishing string. The bowl may be fitted with different types of equipment to grasp the fish and different guides to help center the fish beneath the tool.

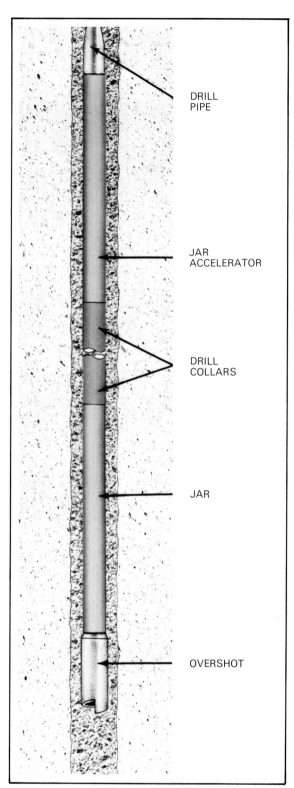

Figure 3.16. Simple fishing assembly

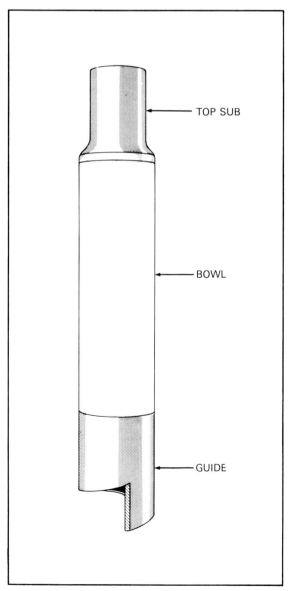

Figure 3.17. Overshot

The external diameter of the overshot should be close to hole size and the grapples close to that of the fish. If the diameter of the fish is within 1/8 inch of the maximum catch size for the overshot, a *spiral grapple* (fig. 3.18) is used. If the fish diameter is well below the maximum catch size—say, 1/2 inch—a *basket grapple* assembly (fig. 3.19) is installed. A basket grapple assembly made up with a mill packer can be used to dress the top of a mildly distorted or burred fish so that it can be caught firmly by the grapples. Both types of packers seal around the fish so that drilling fluid can be pumped down through it to clean out the bottom of the hole.

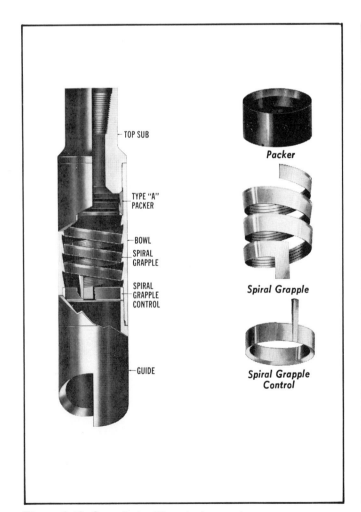

Figure 3.18. Overshot with spiral grapple

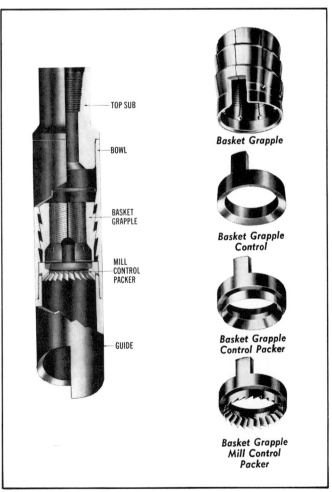

Figure 3.19. Overshot with basket grapple

The driller runs the fishing string to within a few feet of the top of the fish. He starts circulating to clean cuttings and settlings off the top of the fish and to clean out filter cake that may have accumulated inside the overshot. He lowers the fishing string slowly to touch the top of the fish and establish its exact depth. When the fish has been tagged, hook load decreases; the position is marked on the kelly. The string is raised and, with slow rotation to the right, lowered slowly without circulation. If the overshot is centered over the fish, the lowering and right-hand rotation of the string forces the grapple upward within the tapered helix of the bowl, allowing the grapple to expand and the fish to enter the overshot (fig. 3.20). The mark on the kelly is lowered the measured distance from the bottom of the overshot to the inside stop. After the string has been lowered the measured distance, the weight indicator should register a decrease.

Once the fish is engaged, the driller stops rotation and relieves all torque in the string. Then he takes an upward strain. This causes the fish to pull the grapple downward and the wickers on the grapple to bite into the fish. If the fish is gripped tightly, the weight indicator will show an increase equal to the weight of the fish. Circulation is started, without rotation, to clean out the hole before the fish is brought to the surface.

To break out a stand of pipe while coming out of the hole, the string should not be rotated, because right-hand rotation will back the fish out of the overshot. When the top of the fish is pulled through the rotary table, the fish is released from the overshot by bumping down against the rotary slips to break the grip of the grapple. Then the fishing string is rotated to the right and gradually raised until the overshot is clear of the fish.

Many fishing jobs take hours of patient lowering, raising, turning, and feeling with the fishing assembly before the fish is caught. Under these conditions, the measurements taken at the surface pay off. They help determine whether the tool is hitting the top of the fish, the grapples are holding, or the top of the fish is being bypassed.

If an overshot of the proper size has been run and the fish has been bypassed, then the top of the fish is probably in a washed-out section or behind an obstruction. The top of the fish may also be so badly damaged that the grapples cannot engage it. After fishing with the overshot, it should be possible to determine the cause of the difficulty and alter the drill string appropriately.

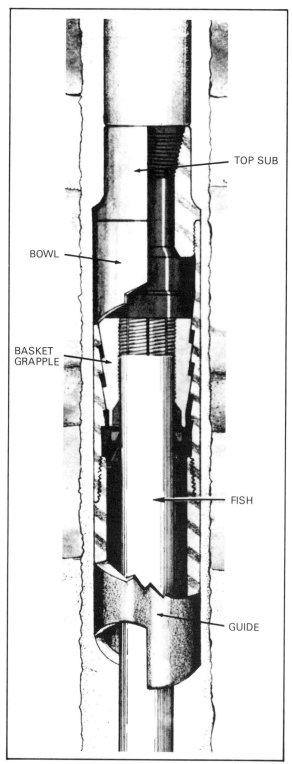

Figure 3.20. Overshot engaging fish

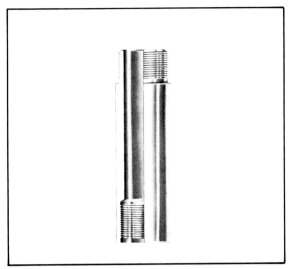

Figure 3.21. Extension sub

If the upper end of the fish is unengageable, an *extension sub* (fig. 3.21) can be installed between the top sub and the bowl of the overshot to allow the damaged top of the fish to go past the grapple. The overshot can then be lowered far enough to engage an undamaged area of the fish.

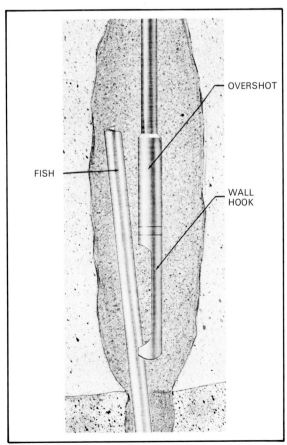

Figure 3.22. Fishing with wall hook in cavity

If the top of the fish is in a washed-out section of the hole, a *wall-hook guide* may be used in place of the regular guide on the bottom of the overshot (fig. 3.22). The distances from the bottom of the guide to the top of the wall-hook opening and from there to the stop in the overshot are measured. The string is run in to a point just above the fish, then lowered slowly with slow rotation until the guide tags the fish. Downward movement is then stopped, but rotation is continued. Torquing up of the fishing string is a sign that the fish is caught in the wall-hook opening. The rotary table is locked and the fishing string is raised. A release of torque signals that the top of the fish has slipped beneath the top of the wall-hook opening and is centered beneath the overshot. To engage the fish, the string is lowered the approximate distance from the top of the wall-hook opening to the stop inside the overshot. If the indicated weight of the fishing string decreases, the fish is caught and can be retrieved.

In a very large washout, a *knuckle joint* may be made up above the overshot to extend the wall hook and overshot out into the cavity (fig. 3.23). Before being run in, the knuckle joint is forced off center by hand to make sure the lip on the wall-hook opening faces about 90° away from the angle of the knuckle joint so that it will contact the fish. Spacer washers between the top sub and the knuckle joint are used to face the opening correctly.

The knuckle joint is kicked off by running in a restriction plug and establishing full circulation. However, good practice is to run the tool in with the restriction plug in place, because the pipe could become wall-stuck in the time it takes the plug to drop to the knuckle joint. The joint is kept straight by running it in without circulation, and then is kicked off in the washout by starting the pumps.

With the knuckle joint bent, the fishing string is rotated slowly to make a sweep around the cavity. If it does not contact the fish, the string is then lowered a few more feet and again rotated. When the wall hook tags the fish, the string torques up and the knuckle joint tends to straighten. The string is slowly raised; when the fish slips into the wall-hook opening, torque is released suddenly. The overshot can then be lowered to engage the fish, and the fish can be removed from the hole.

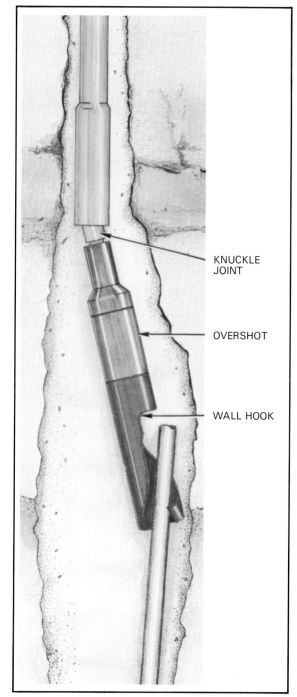

Figure 3.23. Knuckle joint in operation

Fishing for Stuck Pipe

After a fish has been caught in the overshot, the usual procedure is to circulate out the settled cuttings without rotating. If circulation cannot be fully established and the fish cannot be pulled, the fish is almost certainly stuck by cuttings or cavings.

Jarring. The situation described calls for the use of the *jar* and *jar accelerator*. The weight of the drill collars is set down on top of the fish to cock the hydraulic jar and accelerator. Then a moderate pull is taken to stretch the fishing string. This action compresses an inert gas in the jar accelerator above the drill collars (fig. 3.24). At the same time, oil in the hydraulic jar (fig. 3.25) begins to seep between a piston and a sleeve. The sleeve restricts the first few inches of travel, but once the piston is pulled past the sleeve, there is little restriction. The stretch in the drill string, aided by the compressed gas in the jar accelerator, snaps the jar's piston and the drill collar mass upward. The mandrel knocker strikes the knocker sub with great force, jarring the stuck fish upward. Weight is again applied to the string to recock the jar and accelerator for another blow. Jarring is continued until the fish is free and circulation can be restored.

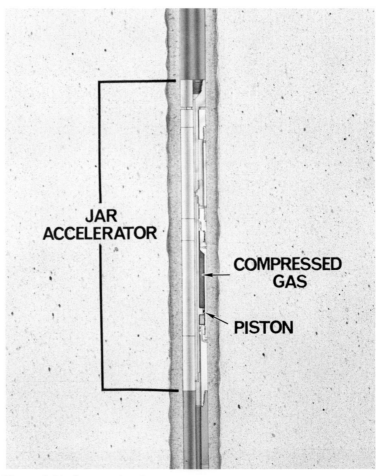

Figure 3.24. Jar accelerator

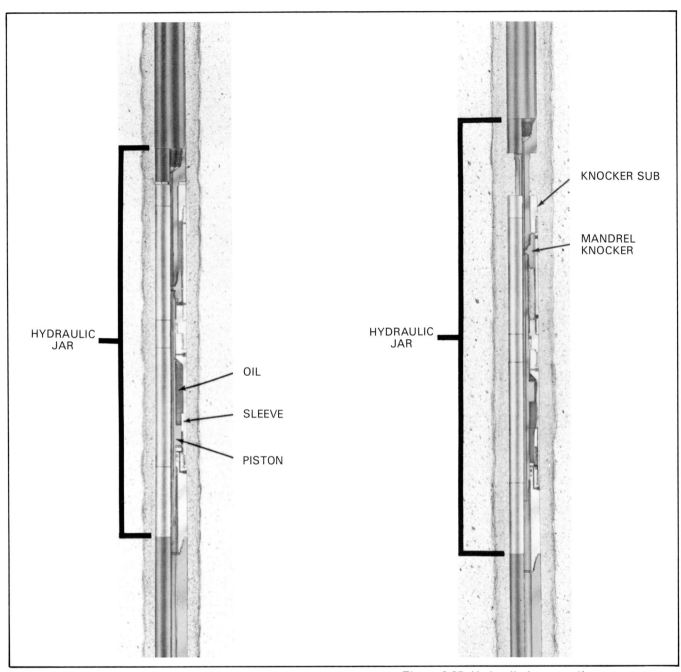

Figure 3.25. Hydraulic jar operation

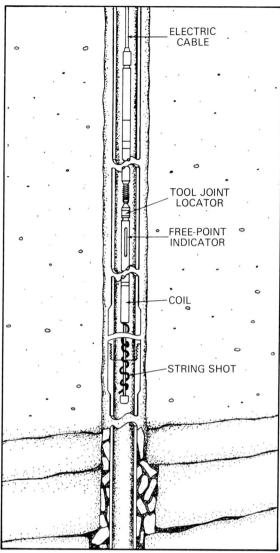

Figure 3.26. Free-point indicator and string shot assembly

Finding the stuck point. If jarring does not free the fish, the next step is to determine at what point in the hole the fish is stuck. The most reliable way to do this is to use a *free-point indicator* (fig. 3.26). This device, consisting of a tool joint locator, an oscillator, and a coil or strain gauge, is run with a *string shot* (a length of explosive primacord) on electric wireline inside the fishing string and into the fish. At intervals, the indicator is stopped, and torque and tension are applied to the drill stem. Stressing the metal affects the electromagnetic field induced by the oscillator and coil; the change is displayed either audibly or visibly on an instrument at the surface. Below the point where the fish is stuck, pulling and turning the drill string has no effect, and therefore no stress is induced, a fact revealed on the surface by the metering device (fig. 3.27).

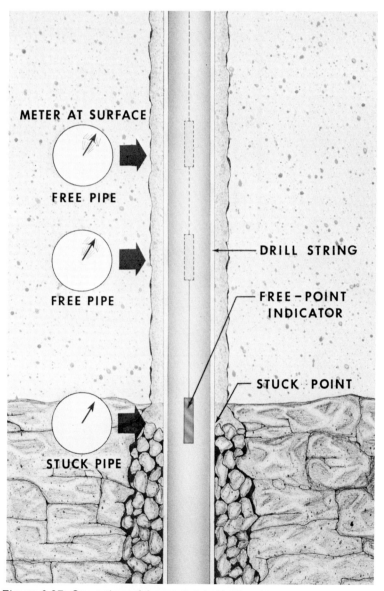

Figure 3.27. Operation of free-point indicator

Backing off. A backoff should be made a few joints above the stuck point so that enough pipe is exposed to guide fishing tools and ensure a clean, undamaged connection. The free-point indicator is raised until the string shot (fig. 3.28) is positioned opposite a tool joint one or two joints above the stuck point. Left-hand torque is applied to the drill stem, and the string shot is detonated to loosen the tool joint. Loss of torque indicates that the joint has been loosened; the backoff is completed by further left-hand rotation. The free-point indicator and the freed section of pipe are then removed from the hole.

The problem now is to fish out the remaining stuck section of pipe. Depending upon the size and type of pipe and how it is stuck, any of several methods may be used.

Washing over. One way to retrieve a stuck fish is to run a washover string (fig. 3.29) to clean debris out of the annulus or to open undergauge hole to full gauge. A typical washover string includes a washover backoff connector, several joints (up to 500 feet) of washover pipe, and a rotary shoe.

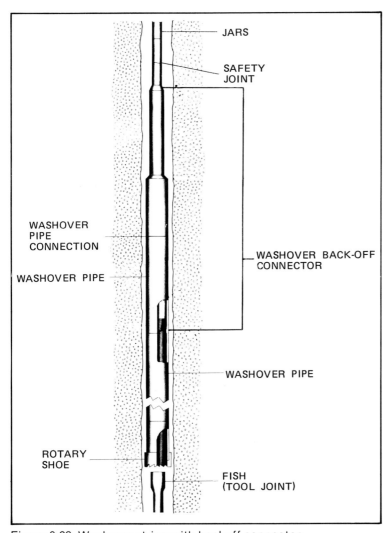

Figure 3.29. Washover string with backoff connector

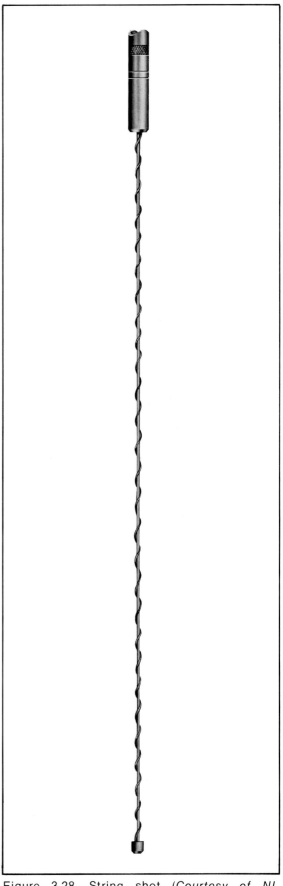

Figure 3.28. String shot (*Courtesy of NL McCullough*)

83

Washover pipe, or *washpipe,* is heavy-walled, N-80 casing with an outside diameter slightly less than that of the hole and an inside diameter large enough to go over the fish. The cutting edge of the washpipe is provided by any of a variety of rotary shoes (fig. 3.30) made of high-grade steel with surfaces of tungsten carbide. Rotary shoes with teeth are generally used to cut formation, while those flat on the bottom (*burn shoes*) are used to cut metal, reduce the diameter or dress the top of a fish, or make a fishing neck on either nearly hole-size round collars or square drill collars.

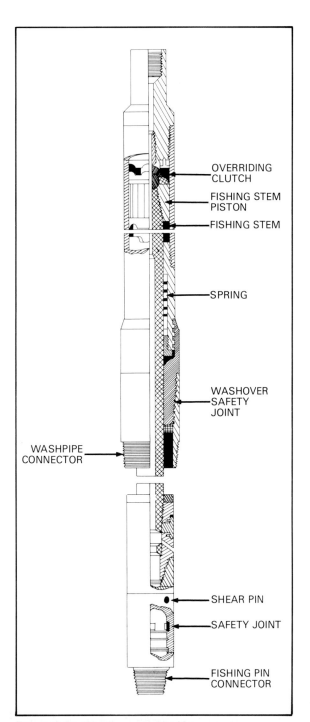

Figure 3.31. Washover backoff connector

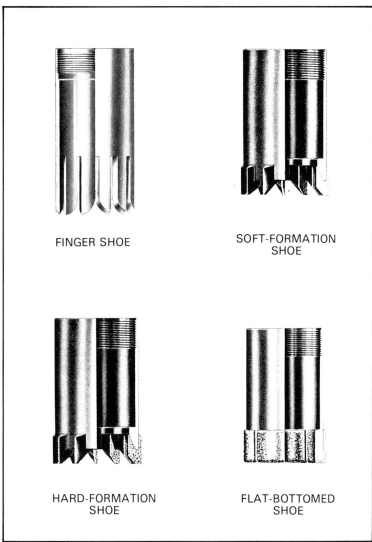

Figure 3.30. Types of rotary shoes

A *washover backoff connector* (fig. 3.31) is designed to engage the top of the fish after the washpipe has washed over and freed part or all of it. The washover backoff connector allows washpipe to be rotated independently of the fish at any time during the operation. If the washpipe sticks, a safety joint permits easy release and recovery of the fishing string, backoff connector, and any washed-over fish from the hole.

The bottom joint of washpipe, made up with an appropriate shoe, is set in the rotary table. Enough washover pipe is added to wash over the fish, but usually not over 500 feet. The washover backoff connector is stabbed down the uppermost joint of washpipe until the threads on the connector tool mate with the washpipe threads. A safety joint, jar, drill collars, and jar accelerator are added; then the assembly is lowered into the hole on drill pipe and run to within 10 feet of the top of the fish. Circulation is established, and the washpipe is lowered slowly over the joint or two of the fish left standing as a guide. Once it is determined that the washpipe is going over the fish, rotation and washing over can begin.

Depending on how tightly the fish is stuck, the washover string should be rotated at 30–50 rpm. Rotation and circulation should be stopped every 20 to 30 feet to check for torque buildup and friction in the washover string. If torque becomes too great, it may be necessary to come out of the hole and remove about half the washpipe. In a crooked hole the top of the fish may be lying under a bend; the shortening of the washover pipe allows it to conform more easily to the curvature of the hole (fig. 3.32).

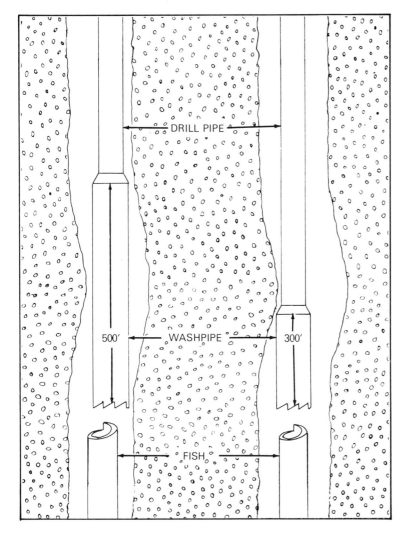

Figure 3.32. Effect of shortening washpipe in a crooked hole

When the entire length of the washpipe has washed over the fish, the fishing-pin connector on the washover backoff connector engages the box of the fish. The overriding clutch allows the washpipe to turn while the fish stands still, but penetration stops—a signal on the surface that the fish is engaged. An upward pull on the fishing string indicates whether the rest of the string below the washpipe is free and can be pulled. If it is not free, a string shot is run down inside the drill string, backoff connector, and fish to within a joint or two of the bottom of the washpipe. The string shot is detonated, the freed pipe backed off, and the fish brought to the surface and stripped out of the washpipe.

Stripping the fish out of the washpipe can be a long, hard job. The drill pipe and part of the fishing string are pulled, broken out, and set back until the washpipe is in the rotary table and the slips are set around it. The washover backoff connector, with the fish attached, is broken out and raised a few feet (fig. 3.33). A split slip holder, or *bowl,* is screwed into the threads in the top of the washpipe, and split slips are set around the fish to hold it. The washover backoff connector is backed out of the fish and laid down. The backed-off fish is stripped out of the washpipe with the elevators. Finally, the washpipe can be pulled if necessary to change the rotary shoe.

To retrieve the remaining stuck fish, the washover backoff connector is again made up on the washpipe and returned to bottom. This sequence of operations (washover, backoff, and pulling) is repeated until all of the fish is retrieved. The last section of the fish is stripped out until the bit is pulled up near the bottom of the washpipe. If the bit is too large to pass through, the fish and washpipe must be pulled, suspended, and broken out together, a complicated and laborious procedure: double stripping 500 feet of washpipe takes about 4½ hours.

Drilling out. Sometimes after backing off above the stuck point and pulling the freed section, the inside bore of the fish still in the hole becomes plugged by settlings and cavings (fig. 3.34), making it impossible to run a free-point

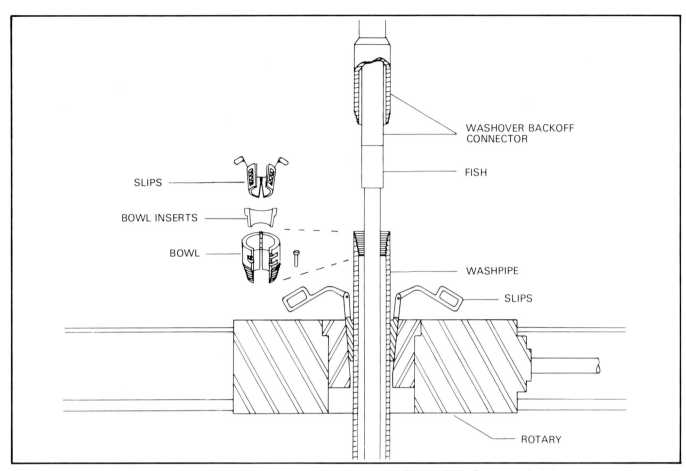

Figure 3.33. Stripping fish from a washpipe using a backoff connector

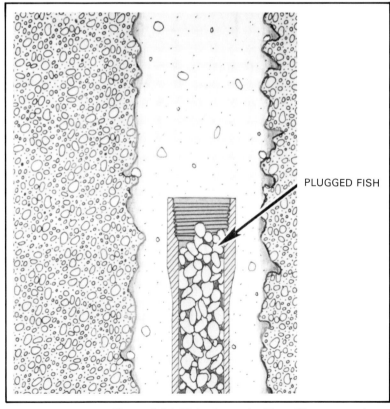

Figure 3.34. Fish plugged with cuttings or cavings

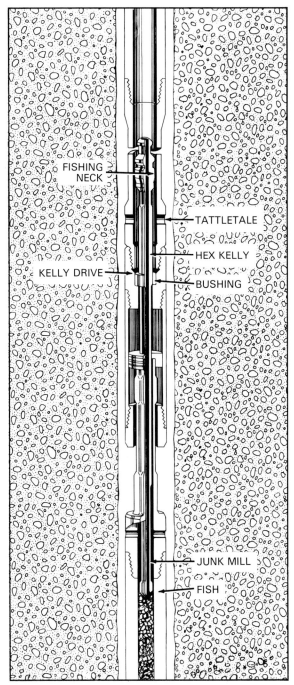

Figure 3.35. Drillout tool

indicator and string shot. Should this happen, a *drillout tool* (fig. 3.35) can be used to reopen the fish. The drillout tool is run to the top of the restricted fish and made up on it.

A special hex kelly tipped with a junk mill is run in on wireline and inserted through a self-aligning bushing built into the kelly drive. Rotating the fishing string then causes the kelly and mill to rotate also. Pump pressure forces the mill into the fish and washes out the cuttings. When the kelly has drilled all the way down, drilling fluid exits through the side opening, or *tattletale*, causing pump pressure to drop.

If there is more debris to be cleaned, the kelly is retrieved by wireline and a short length or *pup joint* of drill pipe made up between kelly and mill. This assembly is run back into the tool, and milling is resumed until the restriction is removed.

Finally, the mill and kelly are again retrieved by wireline, leaving the outer case of the drillout tool attached to the fish (fig. 3.36). The rest of the fish is then removed using normal backoff procedures.

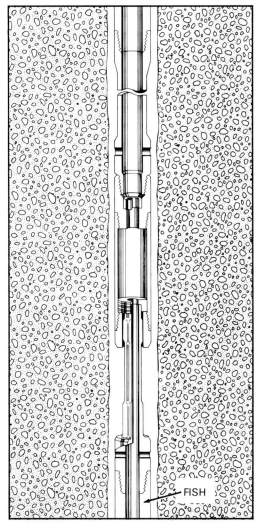

Figure 3.36. Outer case of drillout tool

Cutting pipe. Backing off at a tool joint is usually preferable to cutting pipe because it leaves a clean tool box in good condition "looking up" – that is, facing uphole – so that it can be easily engaged by the fishing string. However, hole obstructions or pipe damage may make it impossible to back off at a connection. If the fish cannot be pulled or jarred loose, cutting the pipe between tool joints may be the only way to retrieve it. Both outside and inside cutters are available for this job.

If for some reason a fishing tool (such as a taper tap) that will not pass a free-point indicator and string shot for a normal backoff is run, an *outside cutting tool* (fig. 3.37) may be run on the bottom of a washpipe string long enough to get below the obstruction. After the cutting tool has been lowered over the fish to the desired depth, it is raised until slips, pawls, or (as in figure 3.37) spring dogs engage a tool joint. The driller takes a strain on the fishing string, causing the knives to move into cutting position. The tool is rotated slowly under strain until the pipe has been cut in two. The washpipe, cutter, and blocked portion of the fish are pulled from the hole. If the lower portion is open, it can be retrieved by washing over without destroying other joints of expensive drill pipe.

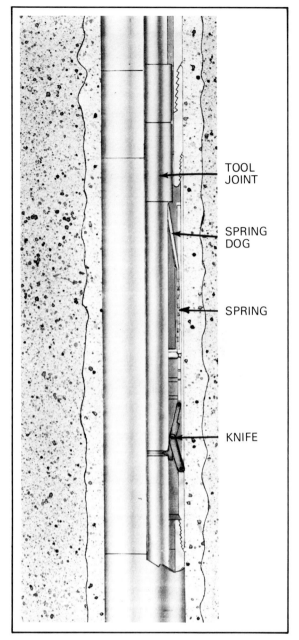

Figure 3.37. Cutting fish with outside cutting tool

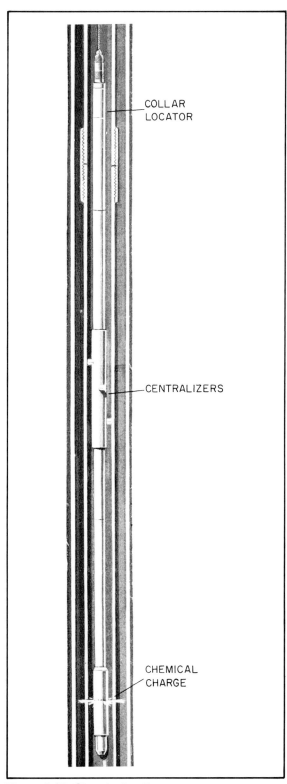

Figure 3.38. Chemical cutter

If the stuck pipe turns freely in the hole, applying torque for a string-shot backoff becomes impossible, and a mechanical cutter may simply engage the fish without cutting it. This situation may call for either a *jet cutter* or a *chemical cutter* (fig. 3.38).

The cutter is lowered inside the fish on a wireline to a depth about one or two joints above the stuck point and activated by electrical impulse from the surface. High-pressure jets of gas or chemicals, forced at high speed and temperature through small openings in the tool, sever the pipe. Because of the close tolerances required for these cutters to work well, they are not reliable in internal-upset drill pipe but are more commonly used to cut tubing.

Freeing Pipe from a Key Seat

Pipe does not become stuck only with the bit on bottom. It can become stuck off bottom during a trip—for instance, in a key seat. Key seating occurs when drill pipe under tension wears a slot into the wall of a crooked hole (fig. 3.7). The slot is usually smaller than the main borehole, too small for drill collars to pass through. If the driller is not careful while tripping out, he can jam the top collar into the key seat so tightly that no amount of weight applied will make it drop free.

To free a fish from a key seat, a free-point indicator and string shot should be run and a backoff made about ten joints above the stuck point so that the top of the fish is in the main borehole and out of the key-seated section (fig. 3.39). The fishing string can then be made up on the top of the fish in open hole.

With an undamaged tool joint looking up, a bumper jar and a key seat wiper may be run and made up on the fish (fig. 3.40). A bumper jar can be used to jar down as well as

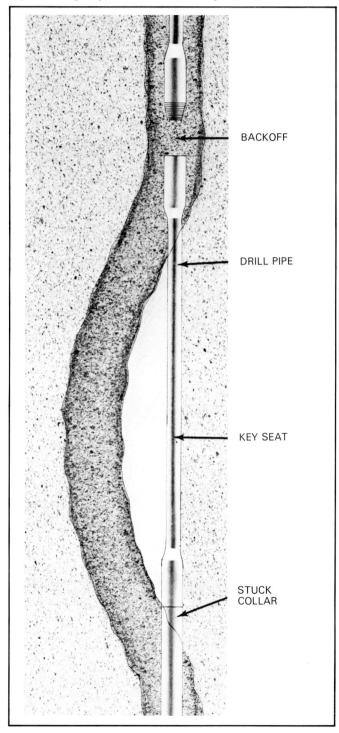

Figure 3.39. Drill collar stuck in key seat

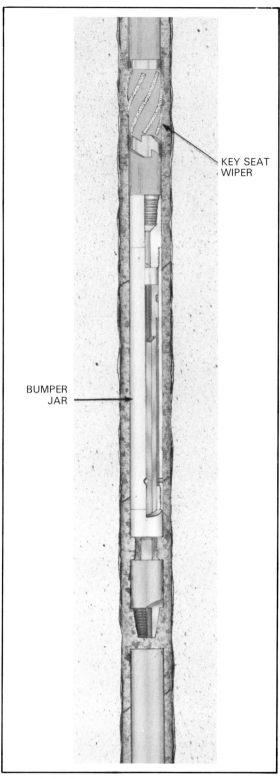

Figure 3.40. Bumper jar and key seat wiper

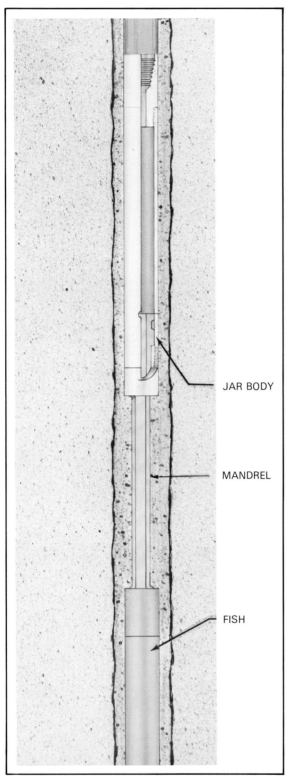

Figure 3.41. Bumper jar prior to downward blow

up. The body of the jar is raised the length of the jar mandrel (fig. 3.41), then suddenly dropped to strike a downward blow. After the fish has been knocked free, the string is lowered and rotated so that the blade reamers on the key seat wiper ream out the key seat. Then the collar string will pass through on the trip out.

If jarring does not knock the fish out of the key seat, a washover may be tried. However, pipe stuck off bottom and not connected to the fishing string would fall to the bottom when freed from the key seat and possibly be damaged or lost. To prevent the fish from falling, an *anchor washpipe spear* (fig. 3.42) is installed inside the washpipe before running in the fishing string.

An anchor washpipe spear is designed to attach to the top of a fish and move up or down inside the washpipe in a controlled fashion. When the fishing string is run into the hole, the anchor washpipe spear, latched inside the bottom joint of washpipe, is made up immediately to the tool joint box of the fish (fig. 3.43). The spear allows the washpipe to travel downhole around the fish as the washover proceeds. However, if the fish begins to fall, the spear's slips engage the inside of the washpipe and prevent the fish from dropping out of the washpipe. Increasing pump pressure allows the fishing string to be raised slowly past the stationary fish to make a kelly connection or free a rotary shoe.

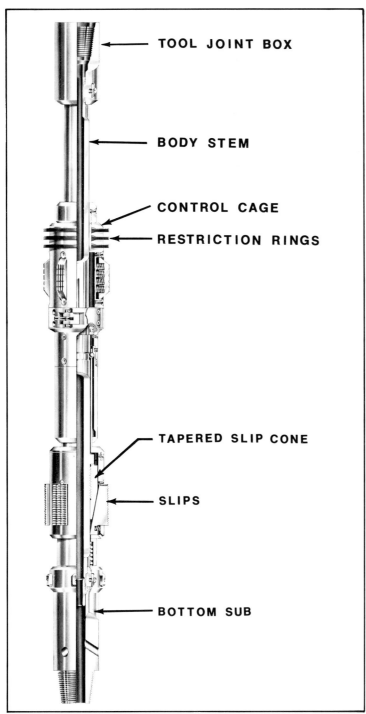

Figure 3.42. Anchor washpipe spear

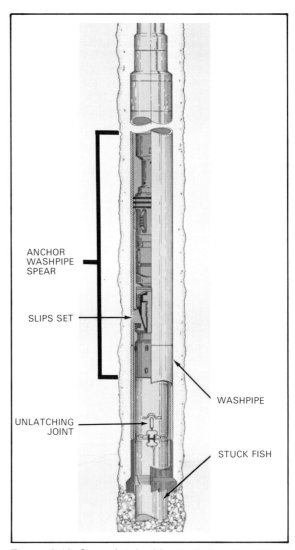

Figure 3.43. Spear latched in washpipe

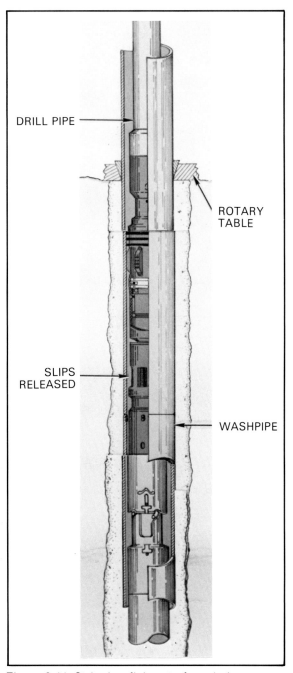

Figure 3.44. Stripping fish out of washpipe

The freed portion of a partially washed-over fish can be recovered by backing off above the stuck point and pulling the washpipe, with spear and fish latched inside, from the hole. Upon reaching the surface, the washpipe is set in the rotary table (fig. 3.44). Drill pipe is run down to and made up on the head of the spear, which is then pulled out of the washpipe with the backed-off portion of the fish. After the fish has been pulled and set back, the spear is again lowered inside the lowest joint of washpipe and latched. The drill pipe is backed out, the washover pipe made up in the fishing string, and the assembly lowered to resume washover operations.

To recover the last portion of the fish from washpipe set in the rotary, the spear slips are released and the spear is lowered and reset in the bottommost joint of washpipe. Then the drill pipe is backed off the spear and the washpipe is pulled until the joint with the spear reaches the surface. With the drill pipe again made up on it, the spear is unlatched by rotation and the spear and fish are retrieved. If the bit is too large to pass through the washpipe, a double stripping job is required.

Freeing Wall-Stuck Pipe

The method used in recovering differentially stuck pipe depends on what the driller was doing when the pipe became stuck. Suppose the driller was fishing out a parted drill string. The top has been engaged by an overshot and jar assembly and circulation has been established, but when an attempt is made to pull up on the fish it is found to be stuck.

Since circulation is not blocked by cuttings or cavings, it is quite likely that the fish is wall-stuck. Differential sticking is related to loss of water from drilling mud into permeable formations, so the first thing that should be done is to cut water loss by changing the mud properties.

With the jar and jar accelerator already in the fishing string, an attempt can be started immediately to knock the fish free. If jarring alone does not free the fish, a slug of oil or other chemicals can be *spotted* around the stuck portion by pumping it down the drill string and up the annulus (fig. 3.45). The spotting fluid penetrates the wall cake, breaking its grip on the pipe. Jarring can continue between spotting intervals.

If oil spotting and jarring do not free the fish within a certain period, the portion of the drill pipe above the stuck point is backed off and pulled, and washover operations begin. If the drill string became stuck while making a connection and with the bit on bottom, a standard washover is performed; if it became stuck in tripping out, an anchor washpipe spear may be used to keep it from falling downhole.

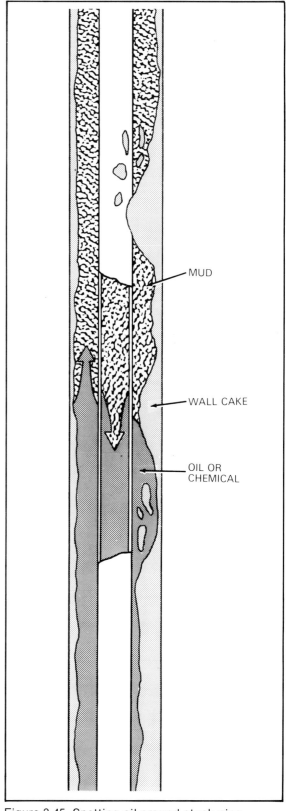

Figure 3.45. Spotting oil around stuck pipe

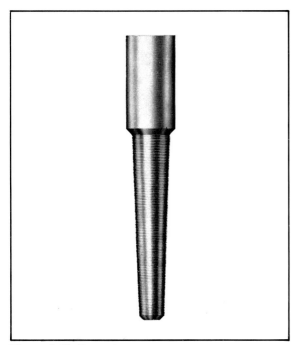

Figure 3.46. Taper tap

Recovering Drill Collars

When a drill collar separates, the break usually occurs at a connection; the pin breaks off in the box or the box breaks off and comes out with the top part of the string. The remaining collars can usually be fished out with a large overshot assembly similar to that used for parted drill pipe. However, if the drill collars are only a fraction of an inch smaller than the wellbore, as in a packed-hole assembly, an overshot may not have enough clearance to go over the collars, and an inside fishing tool may have to be used.

The simplest inside fishing tool is the *taper tap* (fig. 3.46). The tap is lowered into the collar bore and slowly rotated to make its own threads as it engages the fish (fig. 3.47). Some taps have open tips, allowing limited circulation for cleaning off the top of the fish; others have small side jets that move the point of the taper tap about to help locate the top of the fish. None allows a free-point indicator to pass. Once the tap is made up in the fish, the fishing string and fish are tripped out.

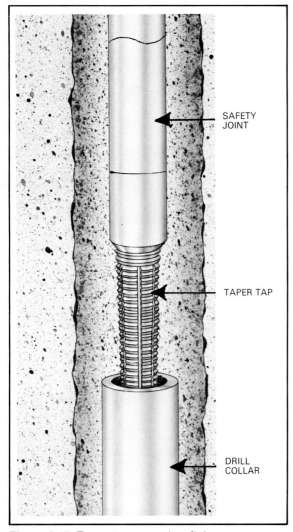

Figure 3.47. Taper tap engaging fish

The taper tap should always be run with a *safety joint* and jar, because once the tap is engaged it cannot be backed out of a stuck collar. If jarring does not free the fish, the safety joint (fig. 3.48) can be broken out by rotating to the left and lowering the drill string. The coarse thread allows the safety joint to be backed out six to eight times faster than a loose tool joint. The box section stays in the hole with the taper tap and the fish. To reattach the fishing string, the pin can be run back into the hole and made up in the box.

Another inside fishing tool is the *releasing spear*. There are many types, one of which is shown in figure 3.49. The spear is made up on the fishing string and lowered, with circulation, to the top of the fish. Circulation is stopped, and the spear is lowered slowly inside the fish until the weight indicator shows a decrease, indicating that the bumper ring has seated on top of the fish. The fishing string is rotated one or two turns to the left. Taking an upward strain then causes the spear to engage the inside of the fish tightly. Circulation can then be resumed and the fish pulled. A sharp downward bump, followed by one to two rounds of right-hand rotation, will disengage the spear if necessary. The spear may also be released by holding a light upward strain and rotating out of the fish.

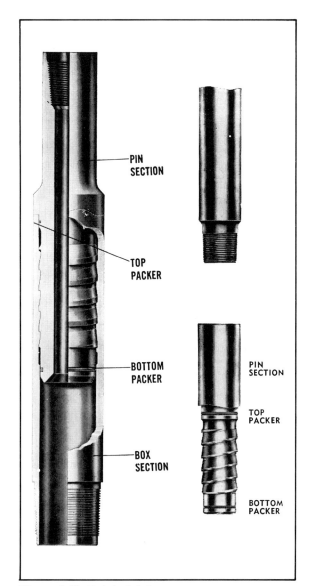

Figure 3.48. Safety joint

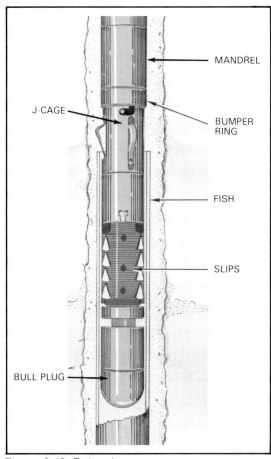

Figure 3.49. Releasing spear

Fishing for Wireline

Wireline tools are commonly used for logging, directional surveying, and other purposes. When a wireline breaks, the line below the break falls and piles up on top of the equipment being run. Before the equipment can be retrieved, the line must either be straightened or be removed from the hole. A proven means of recovering the line is the *rope spear* (fig. 3.50).

The rope spear is made up on the bottom of the fishing string and forced into the mass of coiled line until the line is snagged on its barbs. The rope spear should not be forced too deeply into the line, or the line may ball up and jam the spear in the hole. Another kind of line recovery tool, the *center spear* (fig. 3.51), has two or more prongs with the barbs facing in and is less likely to ball up.

If the equipment on bottom is free, the line is used to pull it out of the hole. If the equipment is stuck, the snagged line is pulled until it parts or pulls out. The rope spear is run as many times as necessary to remove all of the loose line; then the equipment is caught with an overshot.

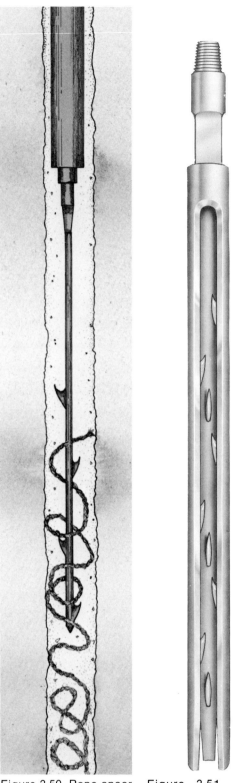

Figure 3.50. Rope spear

Figure 3.51. Prong grab (*Courtesy of Wilson Industries*)

If an instrument becomes stuck with the wireline or cable intact, the *cable-guide method* (fig. 3.52) may be used to recover the cable and instrument. The cable is left attached because it helps hold the fish upright and guides the overshot. First, a light upward strain is taken on the cable to remove any slack. Then a cable hanger is made up on the cable. The cable is cut about 4 feet above the hanger, which rests on the rotary table and prevents the cable from falling into the hole. A spearhead rope socket is made up on the end of the cable extending above the cable hanger. A rope socket, sinker bar, and spearhead overshot are attached to the lower end of the cable left hanging in the derrick and drawn up to the derrickman's position.

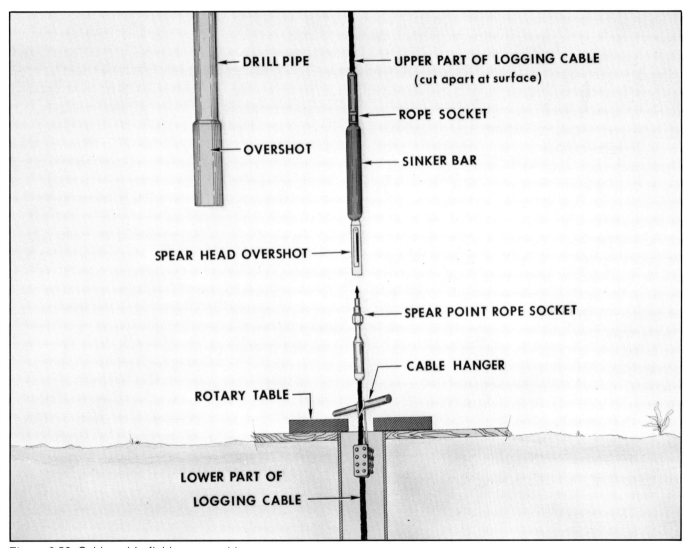

Figure 3.52. Cable-guide fishing assembly

A conventional overshot (to engage the fish in the hole) is made up on the bottom of the first stand of drill pipe. This overshot should be smaller in diameter than a tool joint to keep it from scraping the hole wall and filling with formation debris on the trip in. The drill pipe and overshot are picked up and held over the rotary while the derrickman sends the spearhead assembly down the pipe (fig. 3.53A). The spearhead overshot is attached to the spearhead in the rotary. A light strain is taken on the cable, and the cable hanger is removed. The drill pipe is lowered through the rotary and set on slips (fig. 3.53B).

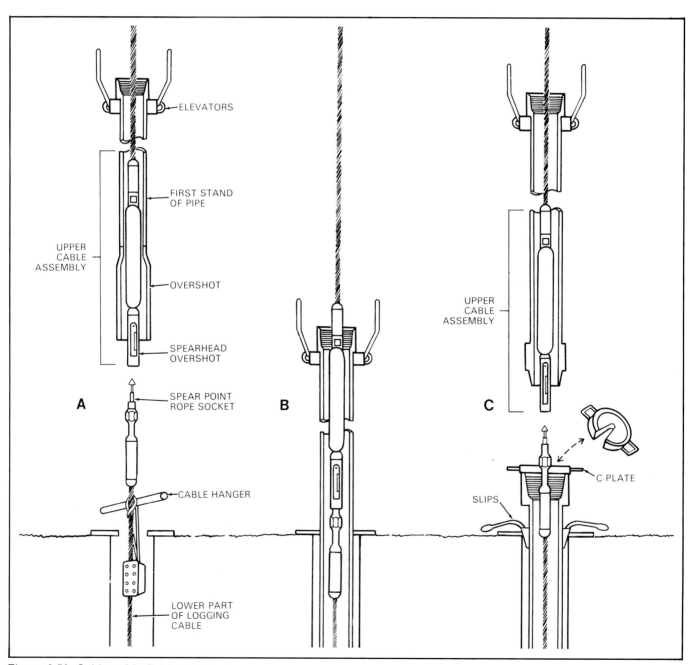

Figure 3.53. Cable-guide fishing method

A C-plate is then placed on top of the drill-pipe tool joint in the rotary to hold the spearhead rope socket. The spearhead overshot is again released and pulled up to the derrickman so that he can thread it down through the next stand of pipe (fig. 3.53C). This procedure is repeated until the overshot on the first stand of pipe engages the fish. If the hole has key seats, the cable will tend to pull into the slot. It may be impossible to run the overshot all the way to the fish without first reaming out the key seat enough to pass the fishing string.

Once the fish is caught by the overshot (fig. 3.54), the cable hanger is again attached and the spearhead rope socket removed. The two ends of the severed cable are secured together with a braided hollow cable called a *snake*. With the elevator latched around the T-bar of the cable hanger, a strain is taken to pull the cable out of the fish. The cable is pulled from the hole and reeled up, and the drill pipe and fish are pulled conventionally.

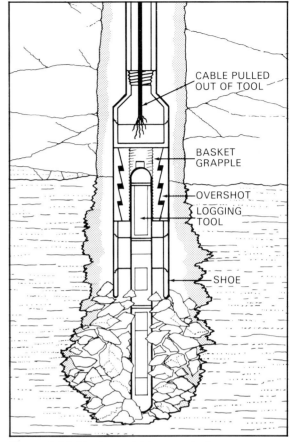

Figure 3.54. Tool caught in overshot

Fishing for Junk

One of the simplest rotary fishing tools is the *finger-type* or *poor boy junk basket* (fig. 3.55). It is run into the hole on the bottom of the drill string to within a few feet of bottom, then lowered over the junk while being slowly rotated. If the basket is nearly hole size, the fingers will gather junk toward the center of the hole and, when weight is applied, bend inward to trap the junk inside. The finger-type junk basket is most effective for a small, solid mass lying loose on the bottom, such as a bit cone.

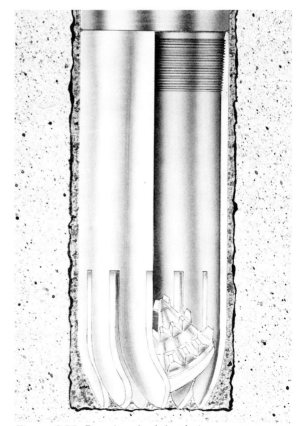

Figure 3.55. Poor boy junk basket

101

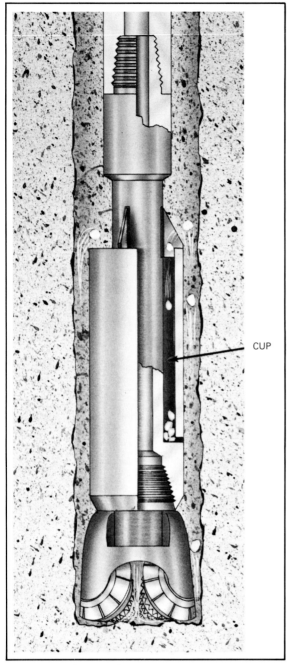

Figure 3.56. Boot sub

A *boot sub* (fig. 3.56) may be run just above the bit during routine drilling to collect small pieces of junk that may damage the bit or interfere with its operation, such as bit bearings or naturally occurring pyrite. Usually, however, a boot sub is run above a mill while it is grinding away a metallic object such as the top of a fish.

During drilling or milling with circulation, the mud flowing upward in the narrow space between the boot sub basket and the hole wall flows rapidly enough to carry pieces of junk with it. When it reaches the annulus above the basket, however, it slows down, and the larger bits of junk drop out into the basket, to be retrieved when the bit or mill is pulled.

The *core-type junk basket* (fig. 3.57) is used to retrieve junk that may or may not be embedded in the formation. A

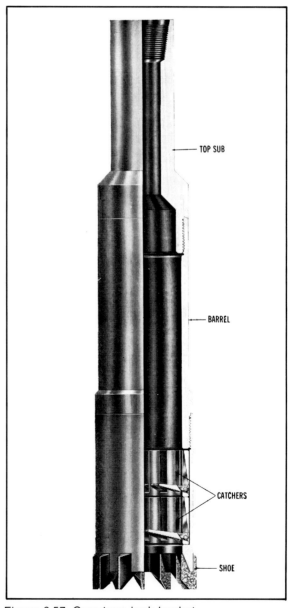

Figure 3.57. Core-type junk basket

mill shoe is made up on the bottom of the tool. After it is run nearly to bottom, mud is circulated at reduced pressure, and the tool is rotated slowly and lowered to contact the junk. Weight is gradually increased. The mill shoe grinds away the protruding edges of the junk, as well as the formation, forcing the junk and a short core into the barrel. Rotation and circulation are stopped, torque is released from the drill string, and an upward strain is taken to break off the core. Upper and lower catchers in the basket hold the core and junk on the trip out. A magnet insert can be used in the tool to pick up small pieces of ferrous metal.

A *reverse-circulation* or *jet-powered junk retriever* (fig. 3.58) uses the hydraulic power of circulating drilling fluid to pick up junk. While being run into the hole, it circulates normally through the bottom to clean cuttings off the top of the fish. A ball dropped down the drill stem then reroutes circulation through jets on the side of the tool and back up through the bottom. Small pieces of debris are carried into the barrel, where folding fingers prevent them from dropping out. If made up with a mill shoe, the jet-powered junk retriever can also cut and catch a core.

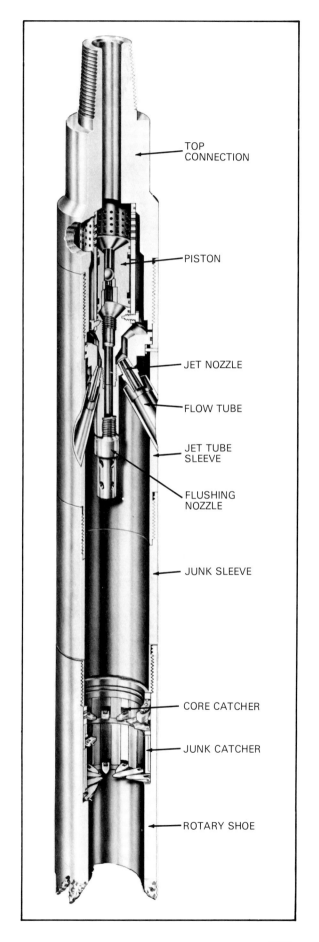

Figure 3.58. Jet-powered junk retriever

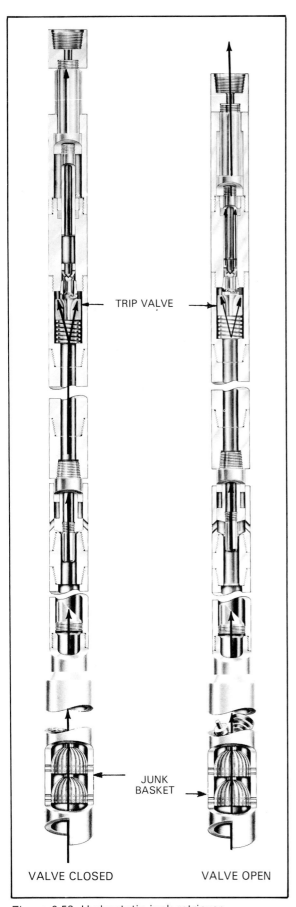

A *hydrostatic junk retriever* (fig. 3.59) also operates hydraulically, much like the jet-powered retriever, but derives its power from a pressure differential. The tool and drill stem are lowered, nearly empty of fluid, into the fluid-filled hole. Once on bottom with the junk basket over the junk, application of weight trips a valve, allowing mud under great hydrostatic pressure to enter the empty drill stem, carrying debris with it. The effect is the same as lowering an empty straw, its top stoppered by a finger, into a glass of water. When the finger is removed, water rushes into the straw. The hydrostatic junk retriever can be opened and closed many times before being removed from the hole.

Ferrous metallic junk can often be retrieved with a *fishing magnet* (fig. 3.60), a powerful permanent magnet with passageways for circulation. A fishing magnet is lowered into the hole with circulation to wash cuttings off the top of the junk. A skirt on the bottom of the magnet keeps the junk from being knocked off during the trip out. Fishing magnets can also be run on wireline, a much faster operation than tripping the drill string in and out.

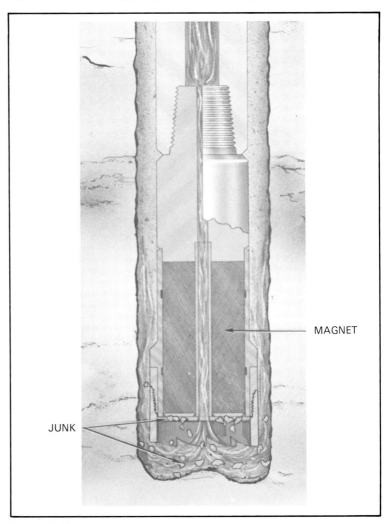

Figure 3.59. Hydrostatic junk retriever

Figure 3.60. Fishing magnet

When junk is so large or oddly shaped that it cannot be readily retrieved with regular junk baskets, a *junk shot* (fig. 3.61) may be used to break it into smaller pieces. This tool contains a shaped charge, an explosive device which focuses the energy of its detonation in one direction. It must be run on drill pipe and collars to keep the force of its explosion from blowing it uphole. The junk shot is lowered on drill pipe and collars until it is just above the junk. When the shaped charge is fired, its downward-directed energy breaks up the junk, which can then be recovered by using a standard junk fishing tool.

Occasionally the force of the explosion embeds the fragments in the formation where they will not impede a roller cone bit, making a separate retrieval run unnecessary. In this case, it is good practice to run a boot sub above the bit to catch the small pieces of junk as they are drilled out of the formation.

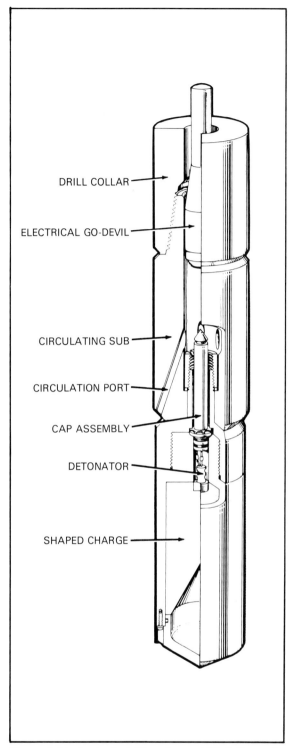

Figure 3.61. Junk shot (*Courtesy of Well Control*)

THE ECONOMICS OF FISHING

Some fishing jobs can go on for months before the fish is retrieved. After a certain period, however, the cost of fishing operations and lost drilling time become prohibitive. As a rule of thumb, once these costs reach about half the cost of sidetracking and redrilling, fishing should be abandoned.

Another way to evaluate fishing economics is to calculate the number of days that should be allowed for fishing, using the following equation:

$$D = \frac{V + C_s}{R + C_d}$$

where
- D = number of days to be allowed for fishing;
- V = replacement value of fish;
- C_s = estimated cost of sidetracking;
- R = daily cost of fishing tool rental and services; and
- C_d = daily rig operating cost.

Suppose a fish worth $150,000 is stuck in a well being drilled at a cost of $5,000 per day. Sidetracking would take an estimated 5 days, so would cost $25,000 plus $20,000 for equipment and cement, or $45,000. Fishing tool rental and services cost $2,500 per day. The number of days allowed for fishing by this method would be as follows:

$$D = \frac{\$150,000 + \$45,000}{\$2,500 + \$5,000}$$

$$= \frac{\$195,000}{\$7,500}$$

$$= 26 \text{ days.}$$

The driller will consider some fishing jobs impractical from the start. For instance, drill collars accidentally cemented in or engulfed in barite are nearly impossible to recover and not worth the cost if they are recovered. The decision is easy to make: start sidetracking immediately. Other decisions are more difficult because the odds of eventual success may not be known.

The cost of sidetracking, however, can be estimated fairly easily. It takes about 5 days to set a cement plug on top of a fish and kick off the hole to bypass it. Knowing the rate of penetration and the length of the original hole he will have to bypass, the driller can estimate the cost of drilling new hole to reach the original total depth.

If the top of the fish is in soft formation, about 100 feet of cement mixed with sand or gravel is run on top of the fish. The hardened cement will deflect a regular bit into the soft formation to start the kickoff (fig. 3.62). In hard formation the cement serves as a base for a deflection tool such as a retrievable whipstock (fig. 3.63).

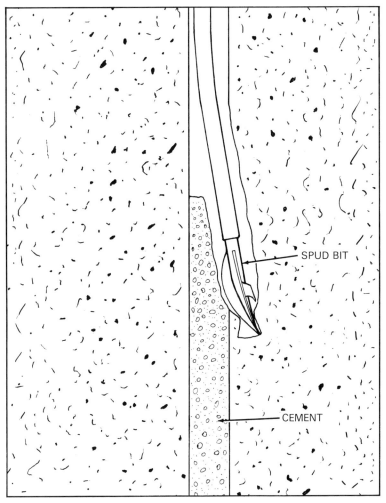

Figure 3.62. Sidetracking in soft formation

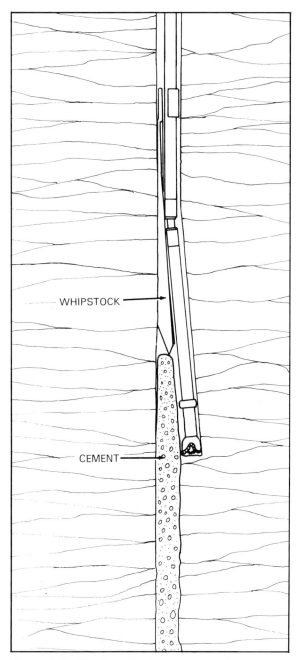

Figure 3.63. Sidetracking in hard formation

Some fishing jobs are easy; some are harder, requiring special tools and the expertise of a fishing specialist. Even then, the best tools and procedures sometimes fail. It is vital to know everything possible about the fish and fishing conditions before starting the job. It is equally important to know when to stop fishing and redrill.

LESSON 3 QUESTIONS

Put the letter for the best answer in the blank before each question.

_____ 1. Twistoffs are usually caused by excessive rotary torque. (T/F)

_____ 2. Surface signs of a twistoff may include –
A. increased drill string weight.
B. increased pump pressure.
C. increased pump speed.
D. increased drilling torque.

_____ 3. Pipe sticking caused by a pressure overbalance is called –
A. sloughing.
B. twistoff.
C. key seating.
D. differential sticking.

_____ 4. If the drill string parts while the bit is on bottom and rotating, the most likely cause is wall sticking. (T/F)

_____ 5. A driller has set back 19 stands (1,710 ft) of drill pipe while tripping out of a 7,260-ft hole. The drill string becomes wall-stuck and parts. The driller then trips out 4,185 ft of pipe above the break. How long is the fish?
A. 1,365 ft
B. 1,710 ft
C. 3,075 ft
D. 4,185 ft

_____ 6. In the previous question, if the fish does not fall, at what depth will the top of the fish be found?
A. 3,075 ft
B. 4,185 ft
C. 5,895 ft
D. 7,260 ft

_____ 7. If an impression block shows the top of a twisted-off section of drill pipe to be badly distorted, what is the fishing tool most likely to be run first?
A. an overshot
B. a string shot
C. a mill
D. a circulating and releasing spear

_____ 8. Which of the following fishing tools engages a fish from the outside?
   A. An overshot
   B. A jar accelerator
   C. A taper tap
   D. A circulating and releasing spear

_____ 9. With its restriction plug in place, a knuckle joint is run into the hole with no circulation. (T/F)

_____ 10. The tool used to locate where pipe is stuck is the—
   A. hydraulic jar.
   B. string shot.
   C. washover backoff connector.
   D. free-point indicator.

_____ 11. The fishing procedure used to free stuck pipe by cleaning debris out of the annulus is called—
   A. milling.
   B. backoff.
   C. washover.
   D. jarring.

_____ 12. In backing the drill string out of a length of stuck pipe, the string shot should be fired—
   A. a joint or two below the stuck point.
   B. as close as possible to the stuck point.
   C. a few joints above the stuck point.
   D. about ten joints below the surface.

_____ 13. A drillout tool is used—
   A. to clean out the bore of a fish plugged with cuttings or cavings.
   B. to enlarge the hole to allow retrieval of a bit.
   C. to obtain a core or retrieve junk embedded in the hole bottom.
   D. to sidetrack a fish.

_____ 14. If possible, it is better to back off a stuck drill string at a connection than to cut it between connections. (T/F)

_____ 15. To cut stuck pipe that turns easily in the hole, _____ can be used.
   A. an outside cutting tool
   B. a jet cutter
   C. a string shot
   D. a key seat

_____ 16. To free drill collars stuck in a key seat during tripping out, the first action should be to—
   A. jar upward.
   B. jar downward.
   C. wash over the drill string.
   D. cut through a drill collar.

_____ 17. When circulation cannot be established through a stuck drill string, which of the following situations is most probably the case?
   A. The drill string is stuck in a key seat.
   B. The drill string is differentially stuck.
   C. The drill string is stuck by a cave-in.
   D. The drill string has washed out (has a hole in it).

_____ 18. A free-point indicator and string shot can be passed through a taper tap. (T/F)

_____ 19. Which of the following equipment would most likely be used to fish out a stuck wireline tool?
   A. Rope spear
   B. Taper tap
   C. Drillout tool
   D. Boot sub

_____ 20. Most washover strings have less than 500 ft of washpipe. (T/F)

_____ 21. If the fish to be washed over is stuck off bottom, which of the following components would the fishing string be likely to include?
   A. A taper tap
   B. A circulating and releasing spear
   C. An anchor washpipe spear
   D. All of the above

_____ 22. The cable-guide method is used –
   A. to recover broken wireline.
   B. to locate the top of a stuck drill collar.
   C. to guide a string shot into a fish.
   D. to recover stuck wireline instruments with the wireline intact.

_____ 23. A poor boy junk basket catches junk –
   A. by drilling through it.
   B. by coring.
   C. by magnetic attraction.
   D. by trapping it inside inward-bending fingers.

_____ 24. A hydrostatic junk retriever –
   A. operates by reverse circulation.
   B. is lowered into the hole with the drill string mostly empty.
   C. is used only in offshore fishing operations.
   D. operates by magnetic attraction.

_____ 25. How many days, at a cost of $2,000 per fishing day, should be spent attempting to recover a $30,000 fish on a rig that costs $4,500 a day to operate, when the estimated cost of sidetracking is $25,000? (Use the formula on p. 106.)
   A. 6⅔
   B. 8½
   C. 12
   D. 27½

As a review exercise for the entire lesson on open-hole fishing, match the descriptions on the left with the terms on the right by placing the letters in the appropriate blanks. Use each letter only once, and only one letter per blank.

_____ 26. Pumping a slug of oil downhole to free wall-stuck pipe

_____ 27. Locates depth at which drill string is stuck

_____ 28. Connects the fishing string to the top of a washed-over fish

_____ 29. Removes debris embedded in bottom of hole

_____ 30. Makes its own threads in the top of a broken-off drill collar

_____ 31. Breaks junk into smaller pieces

_____ 32. Engages broken wireline

_____ 33. Engages the top of a section of stuck pipe from the outside

_____ 34. Uses pressure differential to recover loose junk

_____ 35. Records shape of top of fish

_____ 36. Prevents drill pipe from falling downhole during a washover

_____ 37. Cleans cavings out of the bore of stuck pipe

_____ 38. Self-guiding tool used to dress top of fish

_____ 39. Increases impact while knocking a fish loose

_____ 40. Buildup of mud solids lining the hole

A. Rope spear
B. Hydrostatic junk retriever
C. Drillout tool
D. Oil spotting
E. Jar accelerator
F. Anchor washpipe spear
G. Overshot
H. Core-type junk basket
I. Junk shot
J. Washover backoff connector
K. Taper tap
L. Free-point indicator
M. Wall cake
N. Impression block
O. Piloted mill

Lesson 4
Well Control
(Part I)

Introduction

Well Pressures

Causes of Kicks

Signs of a Kick

Lesson 4
WELL CONTROL, PART I

INTRODUCTION

On January 10, 1901, the Lucas gusher blew in at Spindletop, near Beaumont, Texas. It was one of the most highly publicized of all oilwell blowouts. The Hamill brothers had started the hole 3 months earlier for Captain A. F. Lucas, and 6-inch casing had been set at 880 feet after minor indications of oil. In the next 7 days, the well had been deepened by 140 feet to 1,020 feet, a much faster rate than before. Running in a new bit, the crew had 700 feet of 4-inch drill pipe in the hole when the well started to unload; that is, mud started flowing from the casing. After several hard kicks, well pressure blew the drill pipe out of the hole. Soon a stream of oil and gas was spraying more than 100 feet into the air (fig. 4.1), producing by some estimates 75,000 to 100,000 barrels of oil per day.

Most of the signs of a developing blowout were observable on the Lucas well:

Shows of oil and gas in the mud
Drilling break (faster drilling)
Flow of mud from the well
Pit gain
Lightened hook load

For several days there had been unmistakable signs of oil and gas in the mud, before and after casing was set. These were encouraging, but the highest estimate of production was only 50 barrels a day; no one had any idea that a well could flow at a thousand times that rate. It had taken the contractor nearly 3 months to reach 880 feet, an average of about 10 feet of hole per day. After casing was set at that depth, the drilling rate stepped up to 20 feet daily. The fast progress pleased everyone.

It is now known that bit performance drastically improves when formation pressure and mud column pressure are nearly the same. The Lucas well probably had a bottomhole pressure of 500 pounds per square inch (psi), and the hydrostatic pressure of the mud probably did not exceed 520 psi, assuming mud of 10 pounds per gallon (ppg) in the hole. The driller was able to make hole fast because the well was about to blow out.

As the crew was running drill pipe back to bottom, the well started flowing when mud column hydrostatic pressure became less than formation pressure. The derrickman came down when the first heavy kicks showered the derrick with drilling mud. There was no weight indicator, but the hook load certainly lightened as the pipe began to come out of the well.

Figure 4.1. The Lucas gusher at Spindletop (*Photograph by Frank Trost, January 10, 1901*)

The Hamills had followed accepted practice for cable-tool wells and rotary drilling; there was nothing on top of the casing to shut off an unexpected flow from the well. The three men on the Lucas well were experienced rig hands, but they had never seen or heard of a blowout. When mud and drill pipe started coming out of the hole, they did the only thing they could – they ran to get out of the way of the pipe that was falling around the derrick. The only other action taken that morning was putting out the fire under the boiler.

The well produced from wide-open casing for 9 days before a valve could be attached and closed to stem the flow. Oil was not spouting nearly so high as at first, indicating that the pressure behind the flow had been reduced. About one-half million barrels of crude oil was caught and held behind hastily built earthen dikes. A few days later, however, this oil was lost in a spectacular blaze. Most of those present thought there was a lot more oil underground. They did not realize that the pressure behind the oil was caused by the weight difference between oil and water; when the pressure was gone, the oil was nearly used up.

blowout A *blowout* is an uncontrolled flow of formation fluids or gas from a wellbore into the atmosphere or into lower-pressure subsurface zones. It may begin quietly, with a *kick* – an intrusion of formation fluid causing a pressure imbalance in the hole. In a typical kick, a bubble of gas enters the wellbore downhole, causing an excess flow of mud at the surface. The gas floats up the hole and expands as it nears the surface; finally it blows a slug of fluid out of the well. Quiet follows as another bubble of gas forms and gathers momentum; then another slug of fluid spews from the hole. Each kick is stronger than the one before. When they merge in a continuous rush of fluid, a blowout is underway. The Lucas well was blown clean of mud in a few minutes, and a nearly solid stream of oil followed with no control.

Oilwell blowouts are wasteful, not only of time and money spent to bring them under control, but also of pressure in the formation, which is needed to move the oil from the underground reservoir and raise it to the surface. No one was hurt by the blowout of the Lucas well, but since then many have been killed or injured by similar disasters.

The oil industry has benefited from more than 80 years of rotary rig development and drilling experience since the Spindletop blowout. Scientists have learned how formation pressures occur and can predict the pressures likely to be encountered as wells are drilled deeper. Drillers and engineers have developed procedures for controlling well pressure. Although deep exploration, offshore drilling, and other high-technology drilling practices have increased the hazards of blowouts, experienced crews can minimize the danger by taking the necessary precautions, recognizing the signs of an imminent flow, not panicking when a well kicks, and using the equipment properly.

WELL PRESSURES

Pressures During Routine Drilling

Two basic types of fluids are found in a well: formation fluids and drilling fluids. Drilling fluid is the main line of defense against blowouts. Formation fluids are under pressure; when a formation is penetrated by the drill bit, formation pressure tries to force fluids up the wellbore. The weight of the drilling fluid in the hole exerts pressure against formation fluids, keeping them from entering the wellbore.

If formation pressure exceeds the pressure of the drilling fluid, additional pressure must be imposed to keep formation fluids from rising uphole and pushing out the drilling fluid. This pressure can be imposed in one of two ways—by shutting in the top of the well or by increasing the weight of the drilling fluid. These two procedures are the basis of well control.

Normal formation pressure. Think of the ground as a large open-topped vessel nearly filled with water. The upper surface of the water is called the water table. The top of the container is open because soil and rock are largely porous and permeable; water falling on the ground soaks in and trickles down to the water table.

In normally pressured formations, a connection to the surface is presumed. That is, either the permeable formation communicates with the surface through overlying permeable formations, or the permeable formation outcrops on the surface at the same level as the surface directly above the point where pressure is measured (fig. 4.2).

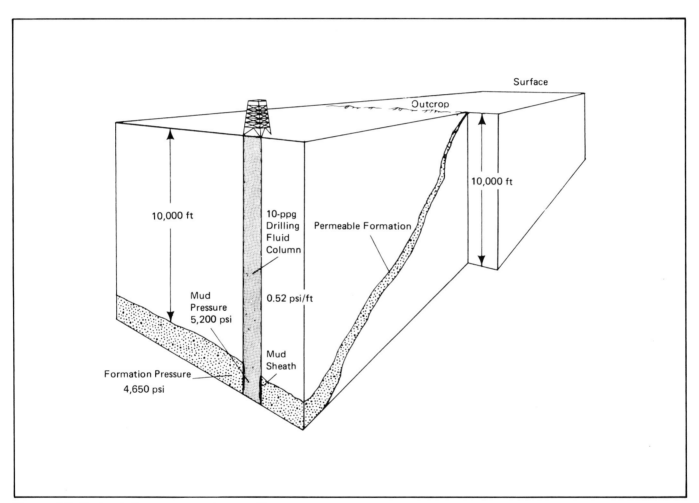

Figure 4.2. Hydrostatic pressures in a wellbore and in a formation

In any vessel, the water at the bottom is under pressure because of the weight of the water above. This pressure – the pressure of fluid at rest – is called *hydrostatic pressure*. Hydrostatic pressure depends upon two factors: the density (weight per unit volume) of the fluid, and its depth. In an open container with a single fluid, the pressure at the bottom is twice as great as the pressure halfway down.

A column of fresh water 1 foot deep and 1 inch square weighs 0.433 pound (fig. 4.3A); that is, it exerts a force of 0.433 pound on the 1-square-inch bottom. Its hydrostatic pressure on bottom is, therefore, 0.433 pounds per square inch (psi). Similarly, a column of fresh water 1 foot deep and 1 foot square (fig. 4.3B) weighs 62.35 pounds, and exerts that force on 144 square inches – again, 0.433 psi. So the hydrostatic pressure of fresh water at a depth of 1 foot, no matter how wide or deep the container, is 0.433 psi.

$$\frac{62.35}{144} = 0.433$$

The *hydrostatic pressure gradient* is the rate at which hydrostatic pressure increases with depth. A hole filled with fresh water will have a hydrostatic pressure gradient of 0.433 psi/ft. The pressure at a depth of 1 ft is therefore 0.433 psi; at 10 ft, 4.33 psi; at 1,000 ft, 433 psi.

pressure gradient

$1 \times 0.433 = 0.433$
$10 \times 0.433 = 4.33$
$1,000 \times 0.433 = 433$

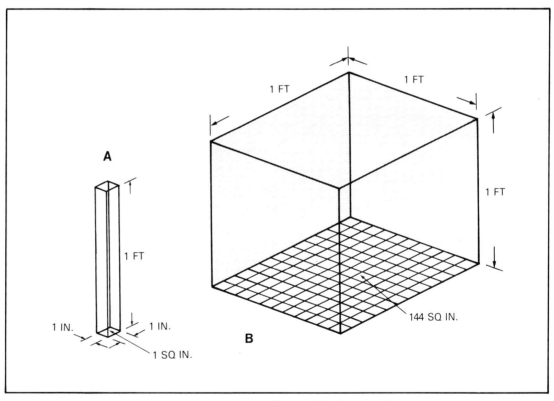

Figure 4.3. Hydrostatic pressure

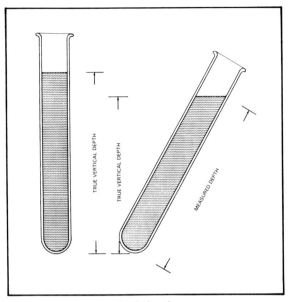

Figure 4.4. True vertical depth

By depth, we mean *true vertical depth (TVD)*. The columns of fluid in figure 4.4 are the same length, but the one on the right is slanted, so its true vertical depth and its hydrostatic pressure on bottom are less than those of the column on the left. Similarly, the shape or width of the container doesn't affect hydrostatic pressure. In figure 4.5, the hydrostatic pressure is the same in all containers at any given depth.

The formula showing how fluid density, TVD, and hydrostatic pressure are related is

$$P = 0.052 \times W \times TVD$$

where
P = hydrostatic pressure (psi);
W = fluid weight or density (ppg); and
TVD = true vertical depth (ft).

Given only its density (in pounds per gallon), the hydrostatic pressure gradient of a fluid, in psi/ft, can be calculated separately by a similar formula—

$$G_p = 0.052 \times W$$

where
G_p = hydrostatic pressure gradient (psi/ft), and
W = fluid weight or density (ppg).

This formula was used to derive the pressure gradients shown in table 4.1. The conversion factor, 0.052, represents the pressure gradient (in psi/ft) of 1-ppg fluid. Fresh water, with a pressure gradient of 0.433 psi/ft, weighs 8.33 ppg.

Figure 4.5. Equal hydrostatic pressures in open-top containers of equal depths

Normal formation pressure varies slightly from area to area. On the Gulf Coast of the U.S.A., the standard gradient is 0.465 psi/ft, that of a column of salt water in the region. Since most petroleum deposits originate in shallow marine sediments, connate water occurring with the petroleum is usually salty or brackish. Therefore, normal pressure gradients are considered to be those between 0.433 psi/ft (fresh water) and 0.465 psi/ft.

Abnormal formation pressure. Formation-fluid pressures not falling between 0.433 and 0.465 psi/ft are generally classified as either subnormal or abnormally high. Subnormal pressures (less than 0.433 psi/ft) may be found in partially or totally depleted reservoir formations, formations at high elevations, and formations that outcrop downhill from a well. Subnormally pressured formations may occur in the same borehole as normally pressured formations.

Abnormally high pressures, too, may be found in boreholes in which normally pressured formations are also exposed. High-pressure zones are usually, but not always, closed formations—that is, not connected to formations with normal pressures. A permeable formation with thick shale formations above and below it will often accumulate fluids squeezed out of the less permeable shale by the weight of the rocks above. These fluids may be under pressure from the entire overburden, the pressure gradient of which is assumed to be about 1 psi/ft. Gradients as high as 0.9 psi/ft have been measured at great depth.

In drilling, such high pressures are usually referred to simply as *abnormal* pressures. Abnormal formation pressures can also be found in permeable formations outcropping at a higher elevation on the surface (as with an artesian water well), in deep or geothermal wells where crustal heat is added to the normal heat of compression, beneath massive salt beds, in the upper parts of large petroleum reservoirs, and in other situations.

TABLE 4.1
FLUID DENSITIES AND PRESSURE GRADIENTS

Pounds per Gallon	Pounds per Cubic Foot	Density (sp. gr.)	Pressure Gradient (psi/ft)
8.00	59.84	0.96	0.416
8.33	62.38	1.00	0.433
9.00	67.32	1.08	0.468
10.00	74.80	1.20	0.520
11.00	82.28	1.32	0.572
12.00	89.76	1.44	0.624
13.00	97.24	1.56	0.676
14.00	104.72	1.68	0.728
15.00	112.20	1.80	0.780
16.00	119.68	1.92	0.832
17.00	127.16	2.04	0.884
18.00	134.64	2.16	0.936
19.00	142.12	2.28	0.988
20.00	149.60	2.40	1.040
21.00	157.08	2.52	1.092
22.00	164.56	2.64	1.144

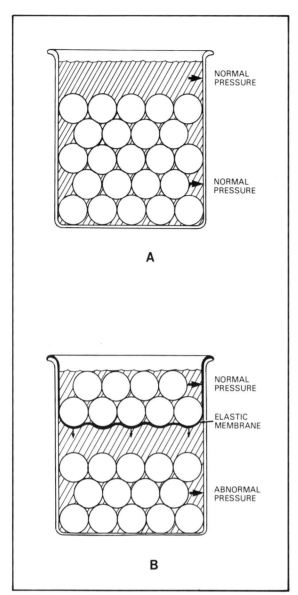

Figure 4.6. Normal and abnormal pressures

A normally pressured reservoir can be compared to a glass full of water and marbles (fig. 4.6A); the marbles represent grains of rock, and the water represents reservoir fluids. The marbles weigh more than the water, but they support themselves—that is, the weight of the marbles on top is borne by those underneath and not by the water. The pressure of the water at any level depends only upon the weight of the water above, not upon the weight of the marbles.

An abnormally pressured formation can be simulated by stretching a sheet of nonporous rubber (to simulate elastic but impermeable shale) over the water in a half-full glass and filling the upper half with marbles and water to the same level as before (fig. 4.6B). Water in the lower half cannot escape through or around the rubber; it supports the weight of both the marbles and the water above it, and therefore has higher pressure than the water in the bottom of the other glass.

Circulating pressure. The term *hydrostatic pressure* applies to fluid at rest—that is, not circulating. If mud is circulated, the hydrostatic pressure is still present, but there is an additional pressure due to fluid friction. The pressure needed to overcome all friction losses in the system is known as *circulating pressure*. Total friction losses, or circulating pressures, vary with the density, viscosity, and gel strength of the mud, length and size of the drill stem, bit jet-nozzle size, annular clearance, and circulating rate. If the pressure of the drilling fluid leaving the annulus is zero, circulating pressure is the same as *pump pressure*.

During drilling, circulating pressure is lost throughout the well, but mostly in the drill stem and bit nozzles. For example, if pump pressure is 2,400 psi (fig. 4.7), the pressure loss due to friction caused by forcing the mud through the drill stem may be 650 psi, leaving 1,750 psi circulating pressure just above the bit. Pressure loss through the bit nozzles may be 1,700 psi, leaving 50 psi to overcome friction through the annulus from bottom to surface.

This friction, the resistance of the annulus to the flow of mud, imposes a back-pressure upon the bottom of the hole that is added to the hydrostatic pressure. The total pressure imposed on the hole bottom in figure 4.7 is 5,250 psi: 5,200 psi hydrostatic pressure plus 50 psi circulating pressure from annular friction.

Total bottomhole pressure is the sum of hydrostatic pressure and all other pressures imposed, such as circulating pressure loss in the annulus or shut-in casing pressure (to be discussed later). This total pressure corresponds to the pressure that would be exerted by a column of mud of a specific weight at that depth. The weight of this theoretical column of mud is known as the *equivalent mud weight*. In the above example, the sum of hydrostatic pressure and imposed (circulating) pressure, 5,250 psi, corresponds to an equivalent mud weight of 10.1 ppg. This value is calculated by using the formula for hydrostatic pressure at a given depth and mud weight:

$$P = 0.052 \times W \times TVD$$

with W representing equivalent mud weight. In the example, $P = 5,250$ psi and $TVD = 10,000$ ft. Substituting these values in the formula gives

$$5,250 = 0.052 \times W \times 10,000$$

$$W = \frac{5,250}{0.052 \times 10,000}$$

$$= 10.1 \text{ ppg.}$$

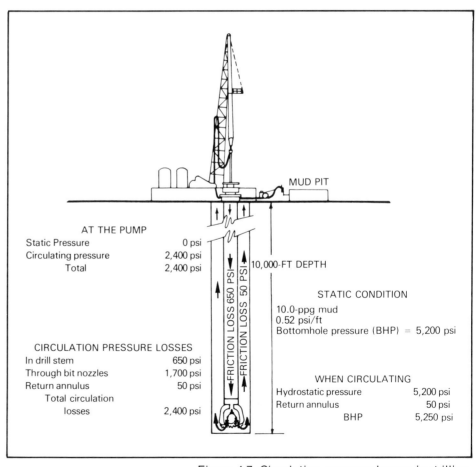

Figure 4.7. Circulating pressure losses in drilling

Pressure balance. Using the 0.465-psi/ft gradient as normal formation pressure, a formation at 10,000 ft would be expected to have a pressure of 4,650 psi (10,000 ft × 0.465 psi/ft). A well drilled with 10-ppg mud, having a gradient of 0.520 psi/ft (table 4.1), would have a hydrostatic pressure of 5,200 psi at that depth. There would be a hydrostatic pressure overbalance, from wellbore to formation, of 550 psi (5,200 − 4,650), and formation fluids would not enter the hole.

On the other hand, if 8-ppg drilling fluid were being used, hydrostatic pressure at the bottom of the hole would be 4,160 psi, 490 psi less than formation pressure. Formation fluids could enter the wellbore. To control normal formation pressure, drilling fluid with a density of at least 9 ppg must be used.

Encountering abnormal formation pressure can have the same effect as drilling underbalanced (that is, with underweight mud). If the bit suddenly entered a 6,500-psi zone at 10,000 ft, there would be a pressure underbalance of 1,300 psi, and 10-ppg drilling mud would not hold back formation fluids. Mud weighing at least 12.5 ppg would be required to control the well. Some of the very deep wells drilled today encounter formation pressures greater than 20,000 psi. Mud weights up to 22 ppg are used to drill these formations.

Suitable downhole hydrostatic pressures can be achieved in various ways. In figure 4.8, column *A* represents a 10,000-ft well filled with 12-ppg mud. Bottomhole hydrostatic pressure is 6,240 psi. Column *B*, with 15-ppg mud, has a bottomhole pressure (BHP) of 7,800 psi.

To calculate the total hydrostatic pressure on the bottom of a hole, the pressures of static fluid columns are added. If fluid is removed, BHP is reduced accordingly. If fluid is added on top of another fluid, total BHP is the sum of the hydrostatic pressures at the bottom of each column.

Column *C*, like *A*, has 12-ppg mud, but the top 1,000 ft is empty. Its hydrostatic pressure at 10,000 ft is 5,616 psi, because it is exerted by only 9,000 ft of fluid: 9,000 × 12 × 0.052 = 5,616 psi.

Column *D* is like column *C* except that 1,000 ft of salt water (pressure gradient 0.442 psi/ft) has been added. Hydrostatic pressure at the bottom of the 1,000 ft of salt water is 442 psi. Downhole pressure at every level in the 9,000-ft mud column is increased by 442 psi, to 6,058 psi at the bottom. The pressure on the bottom would be the same if the salt water were thoroughly mixed with the 12-ppg drilling mud; the mixture would have a pressure gradient of 0.606 psi/ft and would weigh 11.65 ppg (*E*).

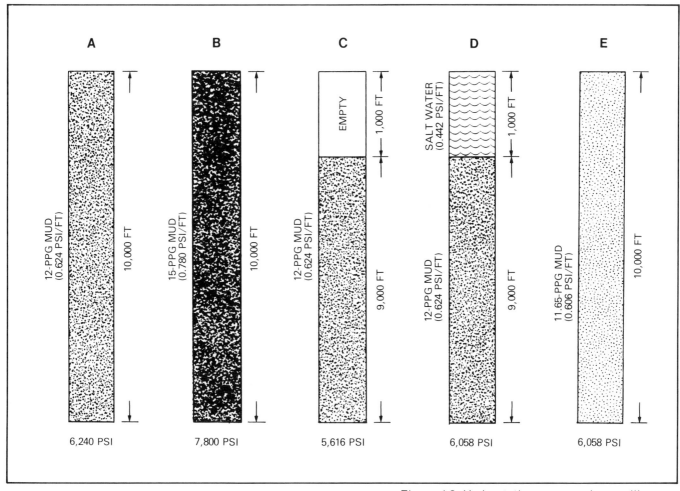

Figure 4.8. Hydrostatic pressures in a wellbore

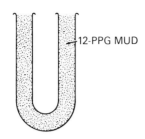

Figure 4.9. Pressure-balanced U-tube

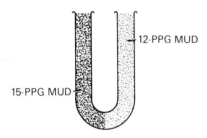

Figure 4.10. U-tube with unbalanced fluid densities

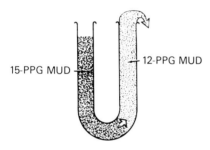

Figure 4.11. Result of density imbalance

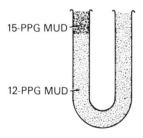

Figure 4.12. U-tube with slug of heavier fluid

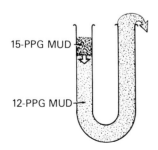

Figure 4.13. Result of heavy fluid slug

Pressures During a Kick

When formation fluids enter the wellbore, the pressure distribution changes. The fluid in the hole is no longer just drilling fluid. Instead, formation fluids, which are almost always less dense, take the place of some of the drilling fluid, which spills out the top of the hole or is otherwise lost from the hole. The weight of the fluid column is reduced. BHP is lowered; it may underbalance formation pressure, allowing further intrusion of formation fluids. To stop the entry of formation fluids downhole, it may be necessary to impose pressure on top of the fluid column by closing the blowout preventer and choke. This action is called *shutting in the well*.

The behavior of fluids in a well can be demonstrated by comparing the well to a U-tube (fig. 4.9). If the left and right legs of the U-tube are filled with fluids of equal density—for example, 12-ppg drilling mud—the weights of the fluids are balanced, and they remain at rest. But if the fluid in the left leg is replaced with heavier fluid, such as 15-ppg mud (fig. 4.10), the heavier fluid will push some of the lighter fluid out of the right leg (fig. 4.11).

When the drill pipe is *slugged* with heavy mud (fig. 4.12), the same thing happens: the heavier mud in the drill stem pushes some of the lighter fluid out of the annulus, causing the fluid level in the drill stem to drop so the pipe can be pulled "dry" (fig. 4.13).

The well is like a U-tube, with the drill stem as one leg and the annulus as the other (fig. 4.14). If the fluid in the U-tube is at rest, the pressure is the same at the bottom of both legs, where they are connected. This point represents the drill bit at the bottom of the hole with no fluid circulating. The bottomhole pressure is the combined hydrostatic pressure of all fluids in each leg, plus any additional pressure imposed on the fluid column, as shown by pressure gauges at the surface.

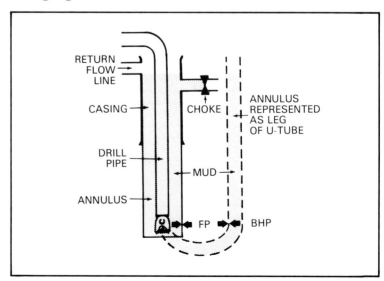

Figure 4.14. The well represented as a U-tube

A freshwater kick. Keeping the U-tube model in mind, look at the well shown schematically in figure 4.15. (This example and the ones following do not represent real-life situations but are for illustration only.) Drilling mud weighing 10 ppg is being circulated down the drill pipe and up the annulus in a 10,000-ft hole. The drill pipe pressure (DPP) gauge shows mud pump pressure, 1,000 psi. Frictional pressure losses in the drill stem, through the bit nozzles, and in the annulus total 1,000 psi, so the mud is under no pressure as it flows through the mud return line into the mud pits. In other words, casing pressure (CP) = 0 psi.

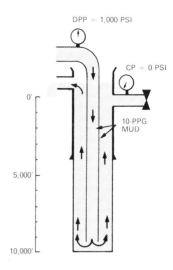

Figure 4.15. A circulating well

If circulation is stopped (fig. 4.16), both drill pipe and annulus are full of 10-ppg mud with a pressure gradient of 0.520 psi/ft. Hydrostatic pressure is 5,200 psi at the bottom. The pressures of the fluids in the drill pipe and the annulus balance each other, so no mud flows from either.

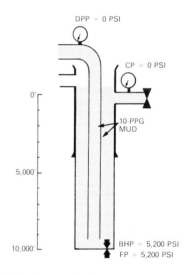

Figure 4.16. Balanced hydrostatic pressures in an open well

If the well is shut in, both *shut-in drill pipe pressure* (SIDPP) and *shut-in casing pressure* (SICP) read 0 psi (fig. 4.17).

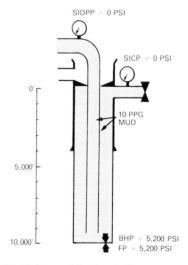

Figure 4.17. Balanced hydrostatic pressures in a shut-in well

127

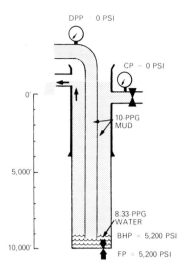

Figure 4.18. Freshwater intrusion, open well

Suppose, however, that there is a 1,000-ft column of fresh water (8.33 ppg) in the annulus. The effect is like slugging the drill pipe with heavier fluid: the fluid column in the annulus is now, on the whole, lighter than that in the drill stem and tends to be pushed out the top of the annulus by the heavier fluid in the drill stem (fig. 4.18).

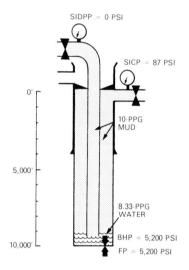

Figure 4.19. Freshwater intrusion shut in

Shutting in the well allows the imbalance to be measured (fig. 4.19). Hydrostatic pressure of the mud at the bottom of the drill stem is still 5,200 psi; but the fluid column in the annulus exerts a hydrostatic pressure of only 5,113 psi: 4,680 psi (9,000 × 0.520) by the mud and 433 psi (1,000 × 0.433) by the fresh water. The result of this imbalance is that the mud in the drill stem exerts 87 psi (5,200 − 5,113) against the bottom of the annular fluid column, forcing it upward.

To prevent fluid from leaving the annulus, a pressure of 87 psi must be imposed on top of the annulus. This is, in effect, what happens when the well is closed in. Simply shutting the blowout preventer and choke adds 87 psi to the 5,113-psi hydrostatic pressure of the fluid in the annulus by opposing the overbalance of the fluid in the drill stem. The added 87 psi, shown as SICP, makes total BHP 5,200 psi. Since no pressure is added to the fluid column in the drill stem, SIDPP is still 0 psi.

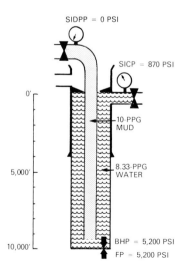

Figure 4.20. Annulus full of fresh water, shut in

If the entire annulus were full of fresh water (fig. 4.20), hydrostatic pressure in the annulus would be 4,330 psi (0.433 × 10,000). The pressure difference, and therefore SICP, would be 870 psi (5,200 − 4,330). The greater the intrusion of formation fluid, the higher will be the pressure that must be dealt with and the harder the well will be to control. So the entry of formation fluids into the wellbore must be stopped as soon as it is noted.

Underbalanced formation pressure. An unexpected encounter with higher formation pressure can affect both SIDPP and SICP. Assume formation pressure of 5,500 psi is exposed at the hole bottom (figure 4.21). The hydrostatic pressure of the fluid column in the drill stem, 5,200 psi, underbalances formation pressure by 300 psi. Formation pressure tends to force fluid out of both the drill stem and the annulus. Keeping the fluid in the well requires an extra 300 psi of imposed pressure at the top of each. SIDPP therefore becomes 300 psi and SICP 387 psi.

Note that the drill pipe serves as a bottomhole pressure gauge. The 300-psi SIDPP reading indicates that formation pressure is 300 psi greater than the hydrostatic pressure of the 10-ppg drilling mud in the drill stem, which is known to be 5,200 psi. The mud weight needed to balance formation pressure can be calculated by using the following formula:

$$MWI = \frac{20 \times SIDPP}{TVD}$$

where
  MWI = mud weight increase (ppg);
  $SIDPP$ = shut-in drill pipe pressure (psi); and
  TVD = true vertical depth (ft).

SICP is not a reliable estimator of formation pressure because it varies with the amount and density of formation fluid in the hole, the height above the hole bottom a gas bubble has migrated, and other factors.

In the situation shown in figure 4.21, mud weight needs to be increased by 0.6 ppg to balance formation pressure:

$$MWI = \frac{20 \times 300}{10,000}$$

$$= \frac{6,000}{10,000}$$

$$= 0.6 \text{ ppg}.$$

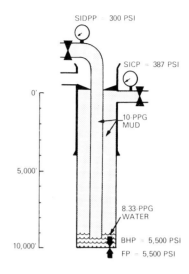

Figure 4.21. Mud overbalanced by formation pressure

A gas kick. A gas kick is harder to handle than a liquid intrusion because gas is compressible. It occupies far less volume under bottomhole pressures than it does at surface pressures. Much lighter than drilling fluid, gas floats up the hole, perhaps 15 to 20 feet per minute. In a well that is not shut in, the closer the gas gets to the surface, the less pressure it is under and the greater its volume.

The relationship between pressure and volume (in an idealized case and under constant temperature) is known as *Boyle's law:*

$$P_1 V_1 = P_2 V_2$$

where
 P_1 = pressure at time 1;
 P_2 = pressure at time 2;
 V_1 = volume at time 1; and
 V_2 = volume at time 2.

A gas kick might have a volume of 10 bbl under normal formation pressure (4,680 psi) at the bottom of a 10,000-ft hole filled with 9-ppg mud. Rising freely to 6,000 ft, it would expand. Hydrostatic pressure at 6,000 ft is 2,808 psi (6,000 × 0.468 psi/ft). Therefore,

$$4{,}680 \times 10 = 2{,}808 \times V_2$$

$$\frac{46{,}800}{2{,}808} = V_2$$

$$V_2 = 16.7 \text{ bbl.}$$

The gas kick would have a volume of 16.7 bbl at 6,000 ft.

In figure 4.22, gas has entered the wellbore at 5,200 psi. The gas intrusion has a volume of 20 bbl and a density of 2 ppg. It forms a 100-foot-long bubble in the annulus. This part of the fluid column contributes only 10 psi of hydrostatic pressure to BHP. Hydrostatic pressure in the annulus underbalances formation pressure by 42 psi (shown as SICP).

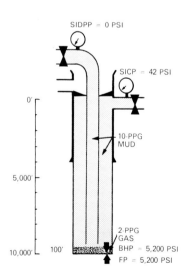

Figure 4.22. Gas kick

Suppose the well is not shut in and the gas kick is allowed to expand freely as it migrates up the annulus (typically at 1,000 to 1,200 ft/h). As soon as it begins to rise and expand, it starts displacing more and more mud from the hole, reducing hydrostatic pressure throughout the annulus. At 5,000 ft (fig. 4.23), the bubble's length has doubled, to 200 ft. However, since its density has been halved (to 1 ppg), it still contributes only its original 10 psi of hydrostatic pressure to the BHP. The 9,800 ft of mud in the hole provides about 5,090 psi, for a total BHP of 5,100 psi. BHP underbalances formation pressure by 100 psi, allowing free entry of formation gas at 10,000 ft.

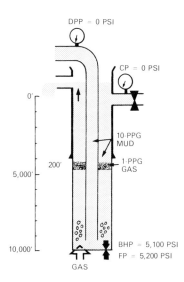

Figure 4.23. Migration of uncontrolled gas kick

In figure 4.24, the base of the bubble has reached 2,500 ft; it is now 400 ft long and displaces 80 bbl of mud. BHP has dropped to 5,000 psi, and gas is entering the wellbore more and more rapidly. The situation is deteriorating quickly, and the well is "coming to see you." (The pressures in this example do not take into account the gas entering downhole, which causes the well to unload even faster than shown.)

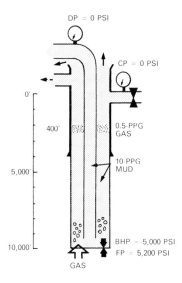

Figure 4.24. Migration of uncontrolled gas kick

What happens to the gas kick if the well is shut in and not circulated? The gas still migrates uphole as before; however, since no fluid has been allowed to escape from the well, the kick occupies the same volume at 5,000 ft as it did at the bottom (fig. 4.25). Therefore, it retains the original formation pressure (5,200 psi). This pressure is exerted in all directions: downward, adding to the hydrostatic pressure of the 5,000 ft of 10-ppg mud below and making BHP 7,800 psi (2,600 + 5,200); upward, overbalancing the 4,900-ft fluid column above by 2,650 psi (5,200 − 2,550), shown as SICP; and outward, perhaps enough to fracture the formation and cause lost circulation. The BHP of 7,800 psi overbalances hydrostatic pressure in the drill stem by 2,600 psi; this difference shows up as SIDPP. (The hydrostatic pressure of the gas itself is negligible and is ignored in this and following examples.)

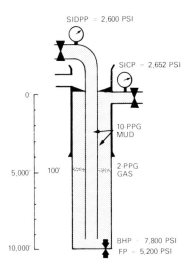

Figure 4.25. Migration of shut-in gas kick

131

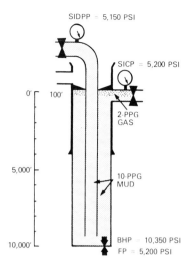

Figure 4.26. Migration of shut-in gas kick

In figure 4.26, the gas bubble has migrated to the top of the annulus, carrying with it the original formation pressure. This 5,200 psi, added to 5,150-psi hydrostatic pressure in the annulus, brings BHP to 10,350 psi, SICP to 5,200 psi, and SIDPP to 5,150 psi. In reality, pressures such as these would never be reached; the formation would fracture, the casing seat would break down, or the BOP would fail first. Leaving a gas kick completely shut in as shown would almost surely end in a blowout.

The behavior of most kicks is more complex than these examples show. Formation fluids mix with drilling fluid, and the top and bottom of a kick may be hard to identify. Rates of uphole migration and pressure change depend upon mud density and viscosity, borehole and drill stem dimensions, whether or not the hole is cased, and what the kick is composed of.

Although SICP is not a reliable estimator of formation pressure, SICP and SIDPP can be used together to determine the density of a kick:

$$D = W - \frac{SICP - SIDPP}{L \times 0.052}$$

where
D = kick fluid density (ppg);
W = mud weight (ppg);
$SICP$ = shut-in casing pressure (psi);
$SIDPP$ = shut-in drill pipe pressure (psi); and
L = vertical length of column of intruding fluid (ft)
   = kick volume (bbl)/annular volume (bbl/ft).

Annular volume, available from tables, depends upon hole size and drill stem dimensions. A 2-bbl kick only 20 ft high in open 10-in. hole might occupy 30 ft in the annulus around drill pipe or 400 ft around drill collars in a packed-hole assembly. Both the density and the height of a fluid column determine its contribution to total hydrostatic pressure.

As an example, suppose a kick has been shut in after causing a pit gain of 5 bbl. SICP is 350, SIDPP is 280, and mud weight is 11.5 ppg. From tables, the annular volume around the drill collars is found to be 0.0218 bbl/ft; that is, 1 vertical foot of hole with drill collars in it can hold 0.0218 bbl of mud. The length of the kick is calculated to be 229 ft (5 ÷ 0.0218). Using the above formula, kick density is calculated as follows:

$$D = 11.5 - \frac{350 - 280}{229 \times 0.052}$$

$$= 11.5 - \frac{70}{11.9}$$

$$= 11.5 - 5.9$$

$$= 5.6 \text{ ppg.}$$

Most kicks are combinations of gas and liquid, with densities ranging from less than 3 ppg (all gas) to 10 ppg (salt water). Between 3 ppg and 9 ppg, the kick may be a combination of gas, oil, and water; the above kick falls near the middle of this range. This type of kick can be expected to migrate uphole faster than a kick with a higher density.

CAUSES OF KICKS

Despite the obvious dangers of abnormal formation pressures, most kicks occur in the course of drilling normally pressured formations. More specifically, half or more of all kicks occur in tripping out or tripping in; and most are due to human error.

Improper Hole Filling

One reason that many kicks occur during trips is failure to keep the hole full of mud. If there is only a small overbalance of formation pressure, measurement of the mud used to fill the hole when the drill stem is removed is critical. This measurement is generally made by using (1) a trip tank measurement, (2) a pump-stroke count, or (3) a pit-level change.

Of the above methods, the trip tank measurement is the most accurate because it shows directly the exact amount of mud needed to fill the hole after a given number of stands of pipe have been pulled. A trip tank is a deep, narrow container that shows several inches of level change for each barrel of fluid. With a gravity-fill tank (fig. 4.27), the hole is automatically filled by mud flowing from the tank as drill pipe is pulled from the well. The trip tank is filled as

Figure 4.27. Gravity-fill trip tank

required from the main circulating tanks. As each stand or row of stands is pulled, the mud level in the trip tank drops by an amount proportional to the volume of steel removed. The exact amount of mud required to fill the hole is compared with the theoretical requirement; the amounts should be identical.

Pump-stroke counts can also be used to measure mud. A reciprocating pump is like a positive-displacement meter. Depending on the size of liners and the length of a stroke, a given number of strokes is required to pump a barrel from the pit into the well. However, this method may not always be reliable. Since no pump is 100% efficient, the number of strokes may cause an overestimate of the amount of mud being pumped. If centrifugal pumps are used to charge the suction, as is commonly done for triplex units, a pump-stroke count may underestimate mud volume because the charging unit may pump through the main pump at low back-pressure.

Pit-level changes can reflect loss or gain of fluid as pipe is pulled from or run into the well, but the level in a large pit does not change noticeably until a fairly large amount is lost or gained. Pit-level instruments should record the average level, provide audible or visual warning to the driller of gains or losses, and record those changes for later reference. However, pit level should not be relied upon as a kick warning because by the time a pit gain is large enough to attract notice, the kick may already be dangerously large.

Because drill collars displace more mud than drill pipe, keeping the hole full while tripping out is most critical when the drill collars reach the surface. The mud level in the well falls four or five times as fast while drill collars are being pulled as when an equal length of drill pipe is being pulled, so the hole should be filled four or five times as often. When a gravity-type trip tank is used, the pulling rate for drill collars must be controlled to keep the hole full.

Pressure Surges and Swabbing

If the drilling mud has high viscosity and gel strength, rapid drill stem movements impose additional pressures on the borehole. The faster the relative movement of mud past the pipe, the greater is the pressure required to overcome annular friction. If the pipe is moving downward through upward-flowing mud, friction is increased, because the mud is flowing past the pipe even faster than past the wall of the hole. The result is a *pressure surge*. The magnitude of a pressure surge depends upon mud weight, viscosity, and gel strength, and upon the relative sizes of drill collars and borehole. A pressure surge can break down a formation, causing lost circulation and leading to an underground blowout. (Formation breakdown and lost circulation are explained in the next section.)

pressure surge

At the beginning of a trip out, bottomhole pressure is reduced because circulation has been stopped. If the bit is pulled out too fast, *swabbing* may occur. Swabbing is a reduction of pressure – the opposite of a pressure surge. Because of friction, mud does not fill in the hole as fast as pipe is pulled out; the mud is sucked uphole, like water in a hand-operated water pump. When the hydrostatic pressure of the mud is only slightly greater than the formation pressure, swabbing may allow formation fluids to enter the borehole. (In fact, swabbing is often used to start a completed well flowing.) Swabbing can also occur if the bit is balled up (covered with accumulated cuttings) and pulled out too fast.

swabbing

The danger of swabbing is greatest with pipe near the bottom, so special care should be taken in starting to trip out. As the pipe is pulled, the volume of mud needed to fill the hole should be carefully noted. If it is less than the theoretical requirement, swabbing has probably occurred. The pipe should be run back to bottom, the contaminated mud circulated out, and the mud conditioned and weighted, if necessary, before making the trip.

Pressure changes due to pipe movement, either surging or swabbing, can be minimized by –

1. proper mud weight to balance formation pressure, plus a trip margin;
2. minimum practical fluid viscosity and gel strength;
3. careful movement of pipe when most of the pipe is in the hole; and
4. adequate clearance between drill collars and hole.

Lost Circulation

Lost circulation (also called *lost returns*) is one of the most serious and expensive difficulties that can happen while a well is being drilled. The ability of exposed formations to withstand the hydrostatic pressure of mud in the wellbore is related to the pressure in its pore spaces and the weight of the overburden. If hydrostatic pressure or a pressure surge in the wellbore exceeds the fracture pressure of a formation, the formation will part, or break down, and mud will be lost from the hole.

formation breakdown

Types of formations to which mud may be lost are (1) coarse and permeable shallow formations, such as loose sand or gravel; (2) *vugular* (cavernous and open-fissured) rocks; and (3) naturally or easily fractured formations. For lost circulation to occur, the rocks must be porous and permeable enough to accept fluid, and wellbore pressure must be great enough to force the fluid into the formation.

Lost circulation is especially hazardous when high-pressure formations are also exposed, because a kick will almost certainly occur when the mud level falls in the well. Conversely, if lost circulation occurs while a kick is being handled with pressure on the preventers, the result may be an underground blowout. The zone of loss must be repaired before surface control can be regained and normal well control procedures resumed.

Sometimes the only sure way to prevent lost circulation is to have enough casing in the well to withstand the hydrostatic pressure needed for formation-pressure control. On the other hand, sometimes a column or slug of heavy mud can be spotted below a zone of loss and topped by a column of lighter mud so that the combined hydrostatic pressures will hold back formation fluids that would otherwise enter the wellbore. Still, the lost circulation zone must be cased and cemented before drilling can continue. Care should be taken to ensure a good cement job; poorly cemented casing has often caused lost circulation.

The pressure at which formations break down has been extensively observed and researched since the mid-1960s. Data on breakdown pressures during squeeze cementing jobs and instances of lost returns have been used to develop correlations among fracture pressure, well depth, and pore pressure for a given area. The graph shown in figure 4.28, for example, was developed for the Louisiana Gulf Coast area by a major operator. The fracture data for normal pore pressure (9 ppg, or 0.468 psi/ft) at various depths are shown as the heavy black line on the extreme left of the family of curves. At 4,000 ft the expected fracture pressure would be equivalent to the hydrostatic pressure of 14.4-ppg mud; that is, the driller would expect the formation to break down under a pressure of 2,995 psi, the pressure of 14.4-ppg mud (pressure gradient,

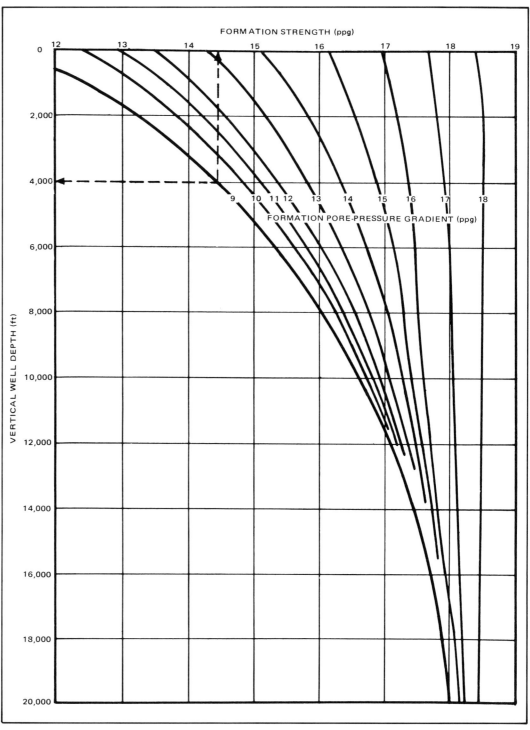

Figure 4.28. Formation strength in the Louisiana Gulf Coast area

0.749 psi/ft) at 4,000 ft. Note, however, that the formations at 8,000 ft could withstand the pressure of 16-ppg mud (pressure gradient, 0.832 psi/ft). Thus formation strength, or resistance to breakdown, increases with depth due to the weight of the overburden.

The other curves show the increased fracture pressure at all depths under abnormal formation pressures. The higher the formation pressure, the more the formation resists the intrusion of drilling fluid.

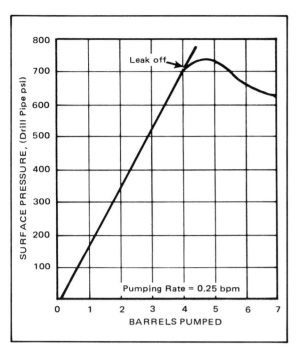

Figure 4.29. Typical leakoff test (surface casing seat at 3,000 ft, mud weight 9.6 ppg)

Formation fracture pressures can be determined in the field by a *leak-off test*. This test is usually made just after drilling 10 to 30 feet through a casing shoe. It measures the maximum mud weight or surface pressure the formation at the casing shoe will withstand before fluid is forced into it. The well is shut in by closing the blowout preventer. Pressure is increased by pumping slowly into the well. At a certain point pressure will begin to drop off, indicating that the exposed formation is taking on significant amounts of mud. The fracture pressure is the total of the surface pumping pressure and the hydrostatic pressure at the casing shoe.

A typical leak-off test is plotted in figure 4.29. The casing seat is at 3,000 ft, and the mud weight is 9.6 ppg. Surface pumping pressure increases in proportion to the volume of mud pumped in (at 0.25 bbl/min) until, after 4 bbl, it reaches about 700 psi. Then the formation begins to fracture and the fluid begins to leak off.

The hydrostatic pressure of 3,000 ft of 9.6-ppg mud is 1,498 psi. This pressure and the surface leak-off pressure of 700 psi establish an observed fracture pressure of 2,198 psi (equivalent mud weight, 14.1 ppg) for the formation at the casing shoe. The expected formation fracture pressure at 3,000 ft, based on data in figure 4.28, is equivalent, at normal pore pressure, to the pressure of 13.9-ppg mud. This value agrees closely with the results of the leak-off test. Were a kick to occur, mud weights up to 14 ppg could be used to control it without fracturing the formation at the casing shoe and causing lost returns.

Abnormal-Pressure Formations

Many blowouts have occurred when abnormal-pressure zones have been penetrated unexpectedly. Such zones can be encountered at almost any depth and require extra care and expense to drill.

One cause of abnormal formation pressure is the compaction of sediments originally deposited as mud, clay, and loose sand. As they are buried under more sediments, the shale is compressed by the weight of the accumulating overburden, becoming denser as excess water is squeezed out. If the permeable formation is cut off from normal pressure, water squeezed out of the shale and into the sand acquires higher pressure and supports part of the weight of the overburden (fig. 4.30). Higher pressures in adjacent formations keep some of the excess water from escaping from the shale; this water remains in the pores of the shale as free water under abnormal pressure. All shales are porous to some degree—although permeability may be practically nil—but shale porosity normally becomes lower upon compaction. Abnormally pressured shales, however, have higher porosity (therefore more water) than normally pressured shales at the same depth.

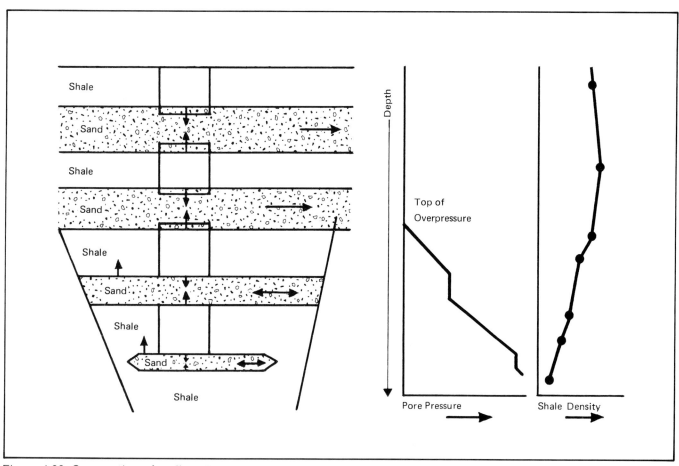

Figure 4.30. Compaction of sediments

Abnormal formation pressures can also occur in formations that outcrop at a higher level than that of the derrick floor (fig. 4.31). This phenomenon is the *artesian* effect. The pressure gradient is normal if TVD is measured from the level of the outcrop; however, since TVD for the well is measured from the rig site, excess pressure corresponds to the height of the outcrop above the well site.

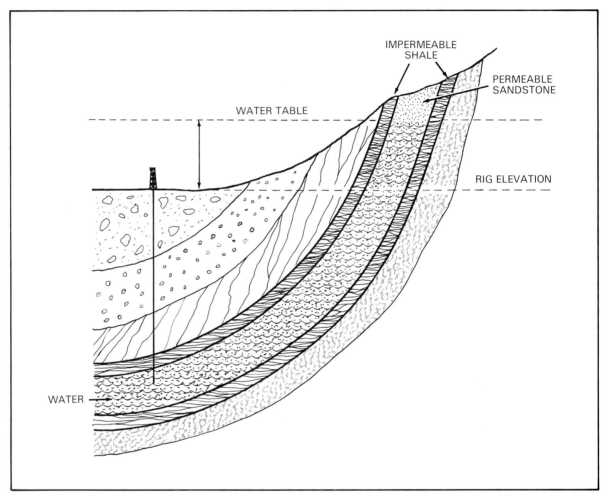

Figure 4.31. Abnormal pressure in a formation outcropping higher than rig elevation

Even a porous, permeable, surface-connected formation outcropping at the same elevation as the rig can have abnormal pressure. In the example shown in figure 4.32, abnormal pressure is found in the gas at the top of the anticline because of the different densities of water, oil, and gas. The gas is abnormally compressed by the weight of the water column, because the water extends higher in the formation than the gas. Since gas is usually found in association with oil and/or water, formations that contain gas are usually abnormally pressured.

Note the similarity between the situation in figure 4.32 and the gas kick in figure 4.22. The top of the anticline corresponds to one leg of a U-tube and the surface outcrop to the other; higher pressures are found in the leg containing the gas than in the leg containing water.

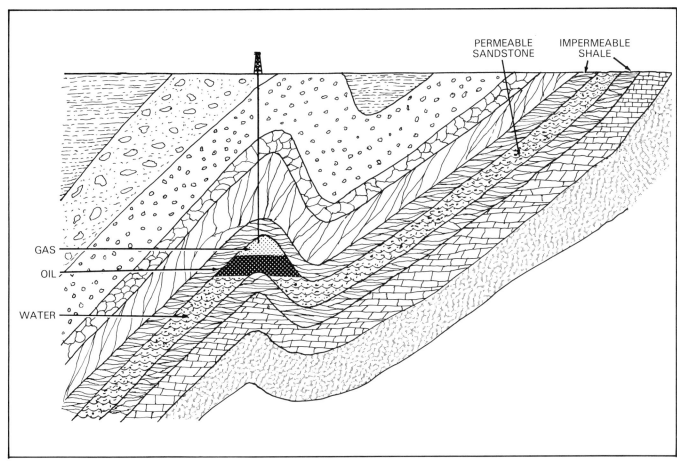

Figure 4.32. Abnormal pressure in a formation outcropping at rig elevation

SIGNS OF A KICK

Some of the preliminary events that may be associated with a kick, not necessarily in order of occurrence, are—

1. pit gain;
2. increased flow of mud from the well;
3. drilling rate changes;
4. decreased circulating pressure;
5. shows of gas, oil, or salt water;
6. rapid variations in rotary torque;
7. fluctuating drill string weight readings; and
8. other indications of abnormal pressure.

Several events may occur in sequence, such as a drilling break, flow of mud from the well, pit gain, and a show of gas, oil, or salt water in the pits. On the other hand, lowering drill pipe too fast may cause formation fracture and lost circulation; the fluid level in the wellbore may drop, exposing a permeable formation and resulting in a kick. In this instance, none of the usual events preliminary to a blowout may occur; the blowout will be underway when the formation fluid hits the surface.

Pit Gain

A gain of fluid in the mud pits is not necessarily the first sign of an impending kick. Unless caused by mechanical switching of fluid in the tanks or similar circumstances, however, it is a positive indication that formation fluid is entering the wellbore.

The size and timing of a pit gain depends upon the type of kick it is associated with. Pit gain can indicate any of several types of downhole situations.

1. A high-permeability shallow gas formation with pressure significantly underbalanced by mud pressure can cause the most dangerous type of kick. It is indicated by rapid flow of mud from the well and fast pit-level rise. The mud shows no sign of gas from the kick source until a large volume of mud flows from the well; flow may start as soon as the high-pressure zone is encountered.

2. A kick from a high-permeability formation slightly underbalanced by mud pressure is difficult to detect quickly. Flow rate is slow at first. Pit gain may be small until the gas is near the surface, but when near-surface gas expansion begins, the mud unloads rapidly, bottomhole pressure is suddenly reduced, and formation fluids begin flowing rapidly into the well. The small pit gain may be accompanied by a drilling break.

3. Low-permeability (tight) formation underbalanced by mud pressure may cause only slow pit gain and, if the underbalance is small, only frothy, gas-cut mud. Rate of penetration may not change. However, it is dangerous to

assume that the formation is tight without thorough knowledge of the specific formation.

4. Kicks following a trip occur when the formation is only slightly overbalanced by the hydrostatic pressure of the mud or when formation pressure is underbalanced but tight, so that swabbing brings gas into the well. A small amount of gas may move part of the way up the hole. When the well is circulated, the gas nears the surface and expands rapidly before circulation from bottom is complete.

If there is even a remote chance of a kick, the drilling rig should have a quick-reading pit-level indicator to show gain or loss of fluid. For an exploration or development well where high formation pressures are expected, an indicating and recording pit-level instrument is required.

When mud is not being circulated, all pit levels in the system are the same. During circulation, however, the level in each successive pit is lower; so the pit-level recorder should show only the average pit level. The recorder should be placed where the driller can see it while drilling and making trips. A gain in pit volume is a signal to the driller for immediate action to check flow from the well and decide whether he needs to close in the well. The driller should be notified whenever mud is added to or taken from a working pit.

Flow of Mud from the Well

The most reliable sign of a kick is a flow of mud from the well with the pumps off. If there is any question that the well might be flowing, the driller should stop drilling, raise the kelly above the rotary, shut down the pump, and check the return line for a flow from the well. This procedure is called a *flow check*. If there is no flow, there is no danger of a kick.

If the well flows with the pump shut down but the pit gain stops when the pump is running, circulating pressure may be providing the only margin of overbalance. The preventer should be closed to check well pressure. If little or no pressure appears on the drill pipe or casing pressure gauges, the weight of the mud may be increased slightly to establish an overbalance of mud hydrostatic pressure over formation pressure. But if pressures appear on the drill pipe and casing when the preventer is closed, well-killing procedures must be begun.

At the beginning of a trip out, a certain amount of mud will tend to cling to the pipe and be pulled out of the hole with it. This phenomenon tends to lower hydrostatic pressure slightly and allow formation fluids to flow into the borehole. The mud return line should be watched carefully to make sure flow does not continue after a reasonable period of adjustment—a minute or so. If it does, a kick is occurring and must be brought under control before the trip out is resumed.

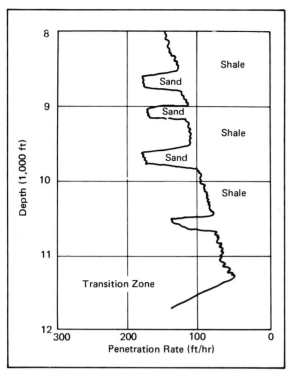

Figure 4.33. Penetration rate vs. hole depth

Drilling Rate Changes

Observing the drilling rate is an indirect means of detecting shale or sand formations containing high pressure (fig. 4.33). Before an overpressured formation is encountered, the usual warning is a gradual but persistent decrease in penetration rate. However, when the cap-rock interval above the abnormal-pressure zone is penetrated, there is often a dramatic increase, perhaps from 10 ft/h to 40 ft/h in as little as 2 or 3 feet of hole. This sudden increase is known as a *drilling break*.

A drilling break usually indicates a formation change, perhaps nothing more than a transition from shale to sand. However, it can also indicate formation pressure approaching or exceeding the hydrostatic pressure of the mud column, especially in soft formations such as coastal shales. The bit drills faster because of the reduction or loss of pressure overbalance; in effect, formation pressure helps break up the rock as the bit fractures it.

In soft formations with a pressure underbalance, the increase in penetration rate may be spectacular—the "bottom falls out of the hole." Usually, however, entering an abnormal-pressure zone causes only a minor change in drilling rate, perhaps from 6 minutes per foot to 5 minutes per foot. An alert driller watches for changes after even a small drilling break. He will be especially alert for flow of mud from the well, for pit gain, and later, when fluid from the bottom reaches the surface, for indications of salt water, oil, or gas in the mud.

All other factors affecting the drilling rate must be held constant to detect a transition from normal to high pressures. Penetration rate is affected by changes in rock types, bottomhole cleaning, circulating fluid properties, rotary speed, and bit weight, type, and condition. The normalized penetration rate can be determined for a given set of conditions, and changes in those conditions can be compensated for. Some mud logging companies determine a normal-rate trend and hold conditions constant or correct for changes, watching for drilling rate increases above the norm to predict high pressure.

Decreased Circulating Pressure

When hydrostatic pressure is the same inside the drill pipe as in the annulus, circulating pressure is directly related to fluid friction losses in the drill pipe, bit nozzles, and annulus. However, if lighter formation fluids enter the annulus, an imbalance occurs. Gas, for instance, rises and expands, displacing some of the heavier drilling fluid; the result is a lighter column of fluid in the annulus than that in the drill pipe. Circulating pressure decreases because the weight of the fluid in the drill string requires less and less help from the pump to push the lightened fluid up the annulus. Unless the pump throttle is adjusted, pump speed gradually increases. If much gas is involved, flow from the well increases, pit gain becomes rapid, and a blowout is underway unless the preventer is closed.

Shows of Formation Fluids

Drilled, swabbed, or trip shows may be circulated from a well after a formation containing them has been penetrated. A show may be obvious, such as a slug of oil or frothy bubbles of gas in the mud; or it may be hard to notice, such as a slight increase in mud salinity. An obvious show may be detected easily by the least-experienced man on the job, but generally an instrument is needed to detect small quantities of gas or chlorides.

A *drilled show* is formation liquid or gas that occupied the pore space of the rock removed by the bit when a section of the hole was drilled. If the formation is thin, or if the hydrostatic pressure of the mud column overbalances formation pressure, little fluid enters the mud stream; if the formation is thick, a drilled show may have considerable volume, especially if it is gas. For this reason, thick gas sands should not be drilled too fast. Drilled shows of gas should be circulated to the surface before cutting another section and adding more gas to further lighten the fluid column.

A *swabbed show* is formation fluid that enters the wellbore due to an underbalance of formation pressure caused by pulling the drill string too fast. It may indicate a small margin of hydrostatic overbalance or a balled-up bit and should warn the driller to be cautious. A *trip show* is another term for a swabbed show but may also include a show that occurs after shutting down the pump but before tripping out the drill string.

Gas shows. Because of its compressibility, gas entering the wellbore can cause serious well control problems. However, a distinction must be made between the small amount of gas involved in gas-cut mud and the much larger and more serious quantity in a gas kick. A gas kick usually begins with a significant pit gain, but without a show of gas at the surface. Gas-cut mud, on the other hand, appears at the surface as frothy mud or mud filled with small bubbles of gas, associated with little if any pit gain.

Gas entering the mud may be any of the following:

1. gas from shale, the so-called high-pressure/low-volume shows that are often associated with thick shale sections;
2. gas from gas-bearing sands that may cause temporary changes in the gas concentration in the mud;
3. *trip gas* that follows round trips of the drill string;
4. *connection gas* associated with shutting down the pump during each drill pipe connection; or
5. gas entering the mud because of insufficient mud weight to control formation pressure.

Since gas is compressible, its presence in the mud may sometimes give the impression of being a more serious problem than it is. A small amount of gas under bottomhole pressure can appear to be a much larger amount at surface pressure. Small amounts of gas mixed into the mud column do not displace much fluid; thus the hydrostatic pressure of the column may not be markedly decreased.

Gas in the pore space of shale feeds steadily into the mud of a drilled hole, forming a baseline on a mud log chart that is often quite predictable. Little attention is usually paid to gas from this source. Gas-bearing sands, however, are more permeable and may cause more serious gas-cutting, especially if drilled too fast. For this reason, many operators drill gas sands more slowly to allow time for show gas to circulate out and prevent reduction of equivalent mud weight.

Trip gas and connection gas are usually the result of lower pressure in the annulus caused by stopping the pump or swabbing. If standard trip and connection practices are followed, an increase in gas-cutting may warn of higher pore pressures.

Saltwater shows. Because salt water is not compressible, saltwater entry is not as serious a well control problem as a gas kick. Small amounts of salt water in the mud cannot be seen, but if 10 barrels or so are swabbed into the well, the mud will appear different when it reaches the surface. An alert man on the ditch can usually catch a sample of formation salt water that is lighter and tests higher in chlorides than the circulating fluid.

Chloride levels of mud entering and leaving the hole should be checked regularly. Brackish and seawater muds are normally high in chlorides, but increased or decreased chloride levels in the returning fluid indicate possible formation water influx. Chloride measurements are usually performed by the mud engineer, who informs the driller.

Oil shows. Shows of oil in the mud are usually more serious than shows of salt water. Gas is usually present in or associated with the oil. Moreover, oil is lighter than water and can lighten the mud enough to underbalance normal formation pressure, leading to a blowout.

Other Indications of Abnormal Pressure

Besides gas-cutting and chloride-level increases, indications of a high-pressure zone may include the following:

1. sloughing shale;
2. shale density changes; and
3. mud temperature increases.

Sloughing shale can be the result of formation pressure in excess of hydrostatic pressure; hydration or swelling of shale by drilling fluid; hole-wall erosion due to fluid circulation, pressure surges, or pipe movement; or fractured or tilted structures. Along with other factors responsible for drilling rate changes, sloughing shale may cause rapid fluctuations in rotary torque and hook load. An otherwise unexplained increase in the size of cuttings at the shale shaker may also indicate sloughing shale.

Shale density normally increases with depth because compaction by the overburden forces the lighter water out of shale pore spaces and into adjacent sands. When the density increase is less than expected, increased formation pore pressures may be expected. In practice, however, an increase in shale density may be hard to detect because it is difficult to select representative particles of shale and measure their density precisely.

Since temperature normally increases with pressure, higher return-mud temperature may indicate higher formation pore pressure. However, it may also result from changes in circulation rate, mud weight, rotary speed, bit weight, torque, or mud chemistry. If changes in these factors can be eliminated or accounted for, a plot of flow-line temperature against depth may warn of abnormal formation pressure (fig. 4.34).

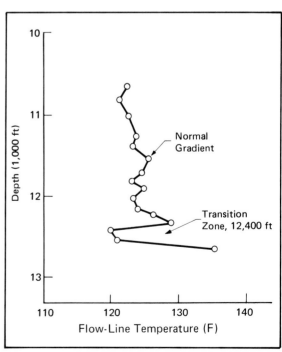

Figure 4.34. Flow-Line temperature vs. hole depth

LESSON 4 QUESTIONS

Put the letter for the best answer in the blank before each question.
Questions 1–34 cover material in **Introduction** and **Well Pressures**.

_____ 1. A kick is –
   A. a blowout.
   B. an intrusion of formation fluid causing a pressure imbalance in the wellbore.
   C. an uncontrolled release of fluid from the well.
   D. a gas fire.

_____ 2. During routine drilling, formation pressure is controlled by –
   A. abnormal formation pressure.
   B. the weight of the drill stem in the hole.
   C. the weight of drilling fluid in the hole.
   D. the BOP stack.

_____ 3. On the U. S. Gulf Coast, normal formation pressure at a depth of 3,500 ft is about –
   A. 1,630 psi.
   B. 3,500 psi.
   C. 350 psi.
   D. 1,870 psi.

_____ 4. At a TVD of 1,000 ft, the hydrostatic pressure of 9-ppg drilling fluid –
   A. is greater in a vertical hole than in a slanted hole.
   B. is greater in a slanted hole than in a vertical hole.
   C. is not affected by hole shape or direction.
   D. depends upon circulating pressure.

_____ 5. The hydrostatic pressure gradient of 11-ppg fluid is –
   A. 0.052 psi/ft.
   B. 0.465 psi/ft.
   C. 0.520 psi/ft.
   D. 0.572 psi/ft.

_____ 6. The vessels shown below contain fresh water. At which point is hydrostatic pressure least?

A   B   C   D

149

_____ 7. Subnormal, normal, and abnormal formation pressures can be found in a single wellbore. (T/F)

_____ 8. Abnormally pressured formations never outcrop at the surface. (T/F)

_____ 9. When fluid is circulated in a well, its hydrostatic pressure drops to zero and is replaced by circulating pressure. (T/F)

(*Questions 10–11*) In a 5,000-ft vertical hole filled with 12-ppg drilling mud, circulating pump pressure is 1,800 psi; pressure loss in the drill stem, 250 psi; through the bit, 1,400 psi.

_____ 10. What is bottomhole pressure (BHP)?
   A. 1,650 psi
   B. 3,270 psi
   C. 1,320 psi
   D. 3,120 psi

_____ 11. What is the equivalent mud weight (EMW)?
   A. 6.9 ppg
   B. 11.4 ppg
   C. 12.6 ppg
   D. 14.5 ppg

_____ 12. A driller using 10-ppg mud to drill through a normally pressured formation at 8,000 ft is drilling underbalanced. (T/F)

(*Questions 13–18*) In the 10,000-ft hole shown at right, what is the hydrostatic pressure –

_____ 13. at the surface?   A. 2,912 psi
                                     B. 3,245 psi
_____ 14. at 2,000 ft?   C. 0 psi
                                     D. 832 psi
_____ 15. at 4,000 ft?   E. 4,160 psi
                                     F. 1,664 psi
_____ 16. at 6,000 ft?   G. 4,090 psi
                                     H. 5,408 psi
_____ 17. at 8,000 ft?

_____ 18. at 10,000 ft?

(*Questions 19–21*) If the hydrostatic pressure of the drilling fluid in the well shown at right exactly balances formation pressure, what is—

_____ 19. BHP?
_____ 20. SIDPP?
_____ 21. SICP?

A. 5,582 psi
B. 86 psi
C. 139 psi
D. 0 psi
E. 5,200 psi
F. 5,720 psi

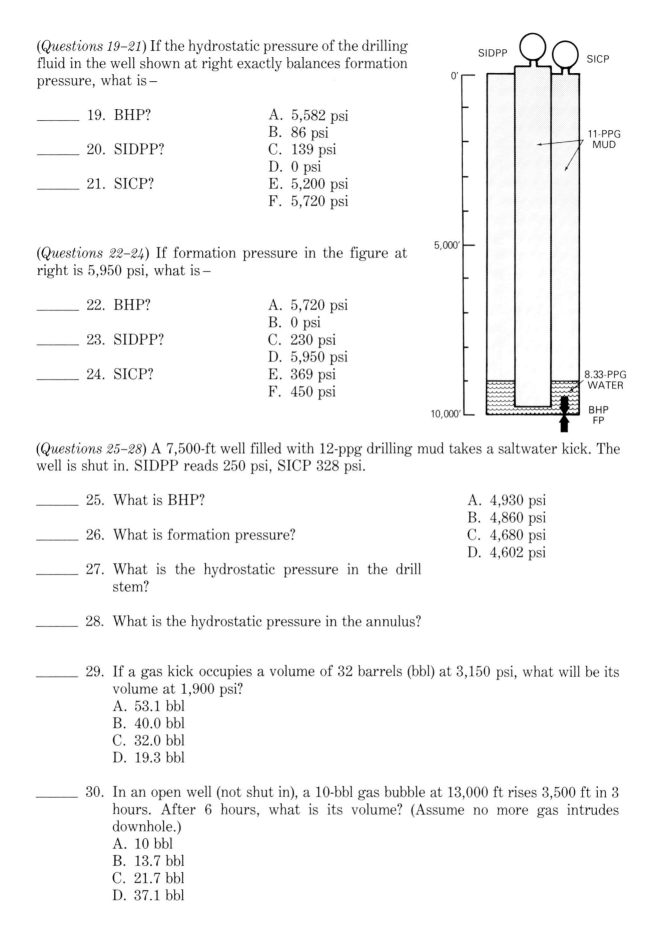

(*Questions 22–24*) If formation pressure in the figure at right is 5,950 psi, what is—

_____ 22. BHP?
_____ 23. SIDPP?
_____ 24. SICP?

A. 5,720 psi
B. 0 psi
C. 230 psi
D. 5,950 psi
E. 369 psi
F. 450 psi

(*Questions 25–28*) A 7,500-ft well filled with 12-ppg drilling mud takes a saltwater kick. The well is shut in. SIDPP reads 250 psi, SICP 328 psi.

_____ 25. What is BHP?

_____ 26. What is formation pressure?

_____ 27. What is the hydrostatic pressure in the drill stem?

_____ 28. What is the hydrostatic pressure in the annulus?

A. 4,930 psi
B. 4,860 psi
C. 4,680 psi
D. 4,602 psi

_____ 29. If a gas kick occupies a volume of 32 barrels (bbl) at 3,150 psi, what will be its volume at 1,900 psi?
A. 53.1 bbl
B. 40.0 bbl
C. 32.0 bbl
D. 19.3 bbl

_____ 30. In an open well (not shut in), a 10-bbl gas bubble at 13,000 ft rises 3,500 ft in 3 hours. After 6 hours, what is its volume? (Assume no more gas intrudes downhole.)
A. 10 bbl
B. 13.7 bbl
C. 21.7 bbl
D. 37.1 bbl

_____ 31. When a gas kick rises in a completely closed-in well—
   A. SICP and SIDPP increase.
   B. SIDPP increases and SICP decreases.
   C. SICP increases and SIDPP decreases.
   D. BHP remains constant.

_____ 32. In figure 4.25, what is the pressure of the gas?
   A. 2,600 psi
   B. 2,652 psi
   C. 5,200 psi
   D. 7,800 psi

_____ 33. In figure 4.26, what is the pressure of the gas?
   A. 0 psi
   B. 5,150 psi
   C. 5,200 psi
   D. 10,350 psi

_____ 34. What is the density of a 120-ft kick in a closed-in well with 13.6-ppg mud and SIDPP = 230 psi, SICP = 290 psi?
   A. 3.1 ppg
   B. 4.0 ppg
   C. 5.8 ppg
   D. 8.5 ppg

Review **Causes of Kicks** and **Signs of a Kick**; then answer questions 35–45.

_____ 35. Most kicks occur when abnormally pressured formations are penetrated. (T/F)

_____ 36. The most accurate means of measuring the mud needed to keep the hole full while tripping out is—
   A. a trip tank.
   B. a pump-stroke count.
   C. pit-level change.
   D. none of the above.

_____ 37. During a trip out, the mud level drops fastest—
   A. at the beginning of the trip.
   B. near the middle of the trip.
   C. near the end of the trip.
   D. after the drill stem is out of the hole.

_____ 38. A reduction of BHP caused by pulling the pipe out of the hole is called—
   A. a surge.
   B. swabbing.
   C. a kick.
   D. annular friction.

_____ 39. Lost circulation can occur –
   A. in naturally fractured formations.
   B. around poorly cemented casing.
   C. when a high-pressure zone is penetrated.
   D. in any of the above situations.

_____ 40. In figure 4.28, what is the formation fracture gradient at 6,000 ft when the formation pore pressure gradient is that of 12-ppg fluid?
   A. 13.5 ppg
   B. 14.4 ppg
   C. 15.3 ppg
   D. 16.0 ppg

_____ 41. The first warning of an impending kick is always pit gain. (T/F)

_____ 42. If mud flows from a well with the pump shut down, but pit gain stops when the pump is started –
   A. the mud pump is defective.
   B. circulating pressure is providing the margin of overbalance.
   C. circulating pressure is greater than mud hydrostatic pressure.
   D. mud hydrostatic pressure is greater than formation pressure.

_____ 43. When the bit approaches an abnormally pressured formation, the penetration rate is most likely to –
   A. increase, then decrease.
   B. decrease, then increase.
   C. decrease continuously.
   D. remain constant.

_____ 44. Formation fluid that occupied the pore space of the rock removed by the bit is called a *drilled show*. (T/F)

_____ 45. Frothy, gas-cut mud is a sure sign of a serious kick. (T/F)

Lesson 5
Well Control
(Part II)

Controlling Kicks

Preventer Drills

Lesson 5
WELL CONTROL, PART II

CONTROLLING KICKS

Once a kick is detected, two things must be done: (1) the kick must be prevented from developing into a blowout, and (2) the conditions that brought about the kick must be corrected. Among the techniques most commonly used to accomplish these goals are–
1. the driller's method;
2. the wait-and-weight method; and
3. the concurrent method.

In practice, combinations of these methods are often used as the situation permits or requires.

The driller who suspects a kick immediately alerts his crew to the problem. He and his crew then shut in the well, using the following procedure:
1. Raise the kelly to clear the tool joint above the rotary.
2. Shut down the pump.
3. Check for well flow.
4. If the well is flowing, immediately open the choke and the choke line manifold, and close the BOP (except on shallow gas kicks).
5. Notify supervisors.
6. Read and record the stabilized shut-in drill pipe pressure (SIDPP).
7. Read and record the stabilized shut-in casing pressure (SICP).
8. Read and record the pit-level increase.
9. Read and record the time.

If a kick occurs while making a trip, driller and crew should take the following steps:
1. Set slips with a tool joint just above the rotary.
2. Install an inside blowout preventer in the drill string and release the valve stem, or close the valve if a drill stem valve is used.
3. Immediately open the choke and the choke line manifold, and close the BOP (except on shallow gas kicks).
4. Install the kelly.
5. Open the drill pipe valve; or pump slowly through the back-pressure valve until it opens, then shut down the pump.
6. Notify supervisors.
7. Read and record SIDPP (or equivalent).
8. Read and record SICP.
9. Read and record the pit-level volume.
10. Read and record the time.

These actions, the first steps in gaining control of a kick, may vary, depending upon equipment and company policy. For example, the procedures outlined are those of a "soft" close—shutting in with the choke line open, then closing the choke. Some operators prefer a "hard" close—shutting the choke before closing the preventers.

The driller and the crew should be familiar with these procedures through drills. Shutting in the well quickly minimizes the amount of formation fluid entering the wellbore and the amount of drilling fluid escaping from the annulus. The smaller the influx, the easier it will be to control. A large influx may exceed the ability of the rig, casing seat, or other exposed formations to contain it.

Shutting in the well also makes it possible to determine, from SIDPP, the formation pressures involved. If the BOP is not closed, there is no accurate way of determining formation pressure, no means of stopping further entry of formation fluid, and no way of calculating the mud weight needed to kill the well. The pressure may build up too high to leave the well shut in, but the situation cannot be studied without at least temporarily shutting in the well.

Figure 5.1 is an example of a worksheet that can be used with most well control methods. Previously recorded information is brought up to date as necessary—for instance, whenever casing is set, mud weight changed, or circulating pressure checked at the reduced pumping rate.

At least once per tour, during routine operations, the pump is run at reduced speed, and both the number of strokes per minute and the reduced circulating pressure are recorded. Some operators run the pump at several speeds to determine the corresponding pressure losses through the drill stem and bit nozzles for the current hole size, depth, and mud weight.

When a kick occurs and the well is shut in, the information in part 2 (SIDPP, SICP, and pit gain) is recorded, along with the time. Pressures do not stabilize immediately when the well is shut in. Formation fluids continue to enter the wellbore until BHP equals formation pressure, so several minutes should elapse before reading SIDPP and SICP. However, even after formation fluids have stopped entering, gas rising through the annulus may cause SIDPP and SICP to increase slowly, so the closed-in period should not be too long. The gauges should be watched carefully after shutting in the well, and readings taken when pressures are reasonably stable.

Fast reaction at the first sign of a kick is the responsibility of the driller and his crew. Once the kick is shut in, however, the toolpusher or the company man usually takes over until the danger is past. Assignment of responsibility depends upon the severity of the kick, contract requirements, and individual company policy.

WELL CONTROL WORK SHEET

Date 2-29 Time 2:00 am

1. **PRERECORDED INFORMATION:**
   Measured depth _10,310_ Depth, TVD _10,000_ Old Mud Wt. _12.0_ ppg
   Casing: Size _OH_ Weight _____ ppf Grade _____ 80% Burst _____
   TVD _____ Ft.
   Drill Pipe _4 1/2_ Weight/ft. _16.60_
   Hole size _8 1/2_
   Drill Pipe Capacity _0.01422_ bbl/ft.
   Pump Output _0.195_ bbl/stk.   Choke line Friction Pressure = _____ psi
                                                       (floating rigs only)

   KILL RATE PRESSURE:
   Pump No. 1 _1000_ psi at _30_ stks/min.
   Pump No. 2 _1000_ psi at _30_ stks/min.

2. **MEASURED:**
   Shut in Drill Pipe Pressure (SIDPP) .... = _200_ psi
   Shut in Casing Pressure (SICP) .... = _325_ psi
   Pit Volume Increase .... = _10_ bbls

3. **DETERMINE INITIAL CIRCULATING PRESSURE (ICP):**
   Kill Rate Pressure (_1000_) + S.I.D.P. Pressure (_200_) = _1200_ psi

4. **CALCULATE MUD WEIGHT INCREASE (MWI):**
   20 X SIDPP (_200_) ÷ DEPTH (TVD) (_10,000_) or Table 4-A = _0.4_ lb/gal
   Add trip margin .... = _____ lb/gal
   Add Old Mud Wt. .... = _12.0_ lb/gal

5. **NEW MUD WEIGHT REQUIRED:** .... = _12.4_ lb/gal

6. **DETERMINE FINAL CIRCULATING PRESSURE (FCP):**
   FCP = Kill Rate Pressure X New Mud Weight ÷ Old Mud Weight
   FCP = _1000_ X _12.4_ ÷ _12.0_ = _1033_ psi

7. **CALCULATE SURFACE TO BIT STROKES (SBS):**
   a. SBS = DP Cap. (bbl/ft) X DEPTH (ft.) ÷ Pump Output (bbl/stks)
   SBS = _0.01422_ X _10,310_ ÷ _0.195_ = _750_ (stks)
   b. Circulating Time to Bit: SBS = (_750_) ÷ stks/min. _30_ = _25_ (min)

8. **GRAPHICAL ANALYSIS:**
   a. Plot final circulating pressure FCP (no. 6) at number of minutes on graph.
   b. Plot initial circulating pressure ICP (no. 3) at the left edge of the graph.
   c. Connect the points (a. & b.) with a straight line.
   d. Across the 10 spaces on the bottom of the graph write in:
      1. Time, surface to bit
      2. Pressure
      3. Surface to bit pump strokes

DRILL PIPE PRESSURE SCHEDULE

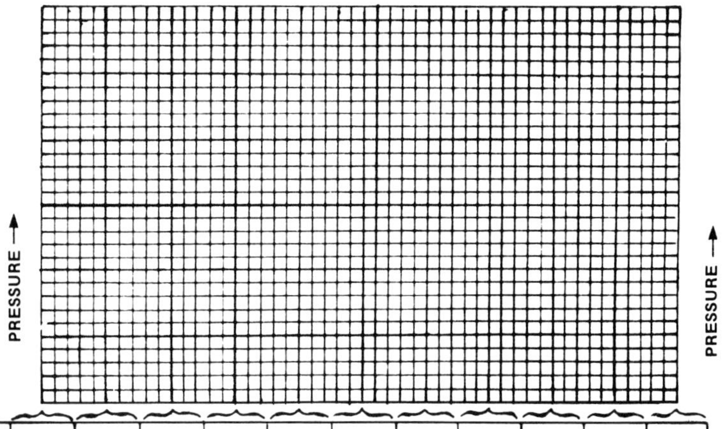

Time										
Press										
Strokes										

MAINTAIN FINAL CIRCULATING PRESSURE TO BOTTOMS UP
AFTER KILL MUD REACHES BIT

Figure 5.1. Well-control worksheet

Blowout Preventers

The *blowout preventer stack* is the primary means of shutting in the well at the surface. Blowout preventer equipment, such as the BOP stack diagrammed in figure 5.2, is designed to—
1. close the top of the hole;
2. control the release of fluids;
3. permit pumping into the hole; and
4. allow movement of drill pipe.

Figure 5.2. Schematic diagram of blowout prevention system (*Courtesy of The British Petroleum Company, p.l.c.*)

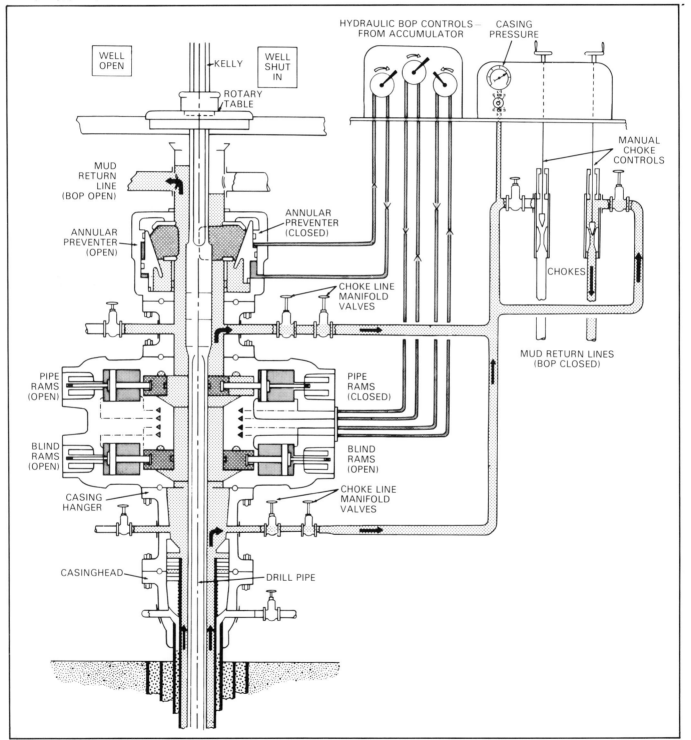

The BOP stack contains two principal types of preventers—ram and annular. The *annular preventer*, usually made up at the top of the stack, has a large ring of steel-reinforced rubber that, when hydraulically compressed, seals the annular space around the drill string or the open hole if there is no pipe in the stack (fig. 5.3). An annular preventer can seal around tools of any shape or size, such as a collar, tool joint, or kelly; in addition, the drill string can be raised or lowered (*stripped*), or rotated with the annular preventer engaged.

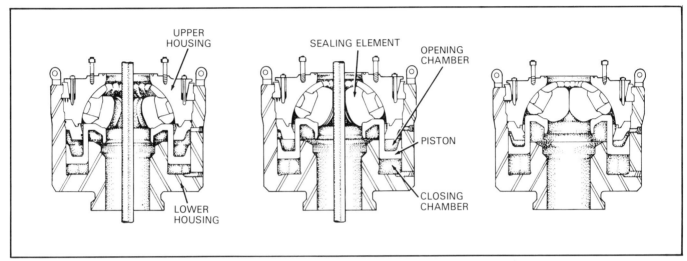

Figure 5.3. Operation of annular blowout preventer (*Courtesy of NL Shaffer*)

Ram preventers, usually mounted below the annular preventer in the BOP stack, operate in pairs by moving from a retracted position clear of the bore into a position where they close off either the annulus or the open hole (fig. 5.4). The rams are made of steel lined with rubber that is forced out under compression to complete the seal. There are three main types:

1. *Pipe rams* close off the annulus around drill pipe, casing, or tubing (fig. 5.5); they have semicircular openings that fit the particular size of pipe in use and center it in the hole. Variable pipe rams are available, which close around a range of sizes – for example, on pipe of 4- to 4½-inch OD.

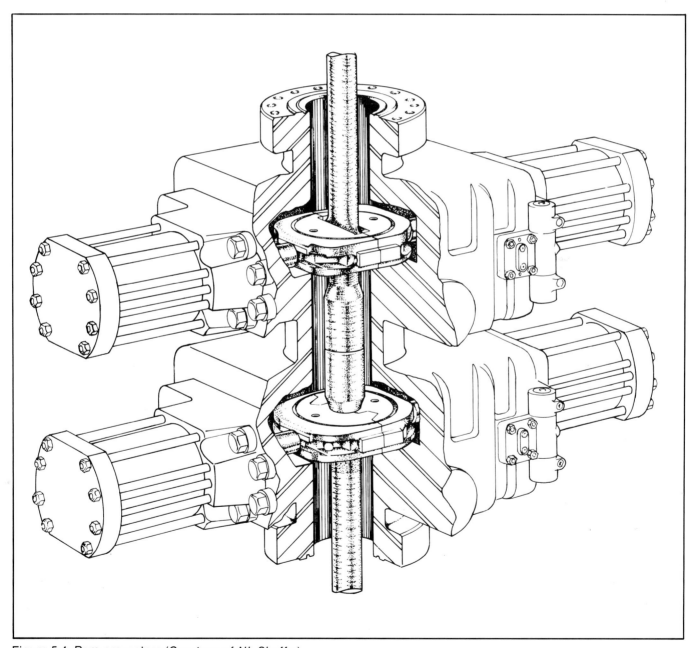

Figure 5.4. Ram preventers (*Courtesy of NL Shaffer*)

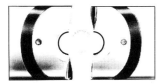  

Rams automatically center pipe in hole.

Self-feeding ram packing seals around pipe before rams are fully closed. Steel retainer plates prevent extrusion of packing.

Generous reserve of packing is fed by additional ram movement, to provide seal after considerable wear.

Figure 5.5. Pipe rams (*Courtesy of Cameron Iron Works*)

2. *Blind rams* are straight-edged rams used only to close open hole (fig. 5.6).

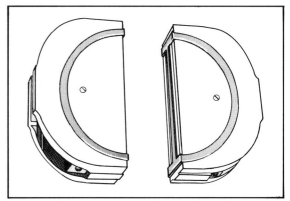

Figure 5.6. Blind rams (*Courtesy of NL Shaffer*)

3. *Shear rams* can be used either to close open hole or to cut through drill pipe and seal off the hole with the drill string in it (fig. 5.7). Shear rams are commonly used offshore, where weather or other factors may make it necessary to move a mobile rig offsite quickly.

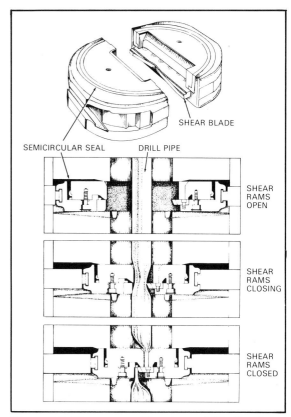

Pipe rams, with their steel blanks and square shoulders, can seal off greater pressures than annular preventers. Closed above a tool joint, they are capable of keeping pipe from being blown out of the hole under extreme pressures.

Figure 5.7. Shear rams (*Courtesy of NL Shaffer*)

A *rotating head* is a special type of blowout preventer used in place of an annular preventer in drilling with pressure at the wellhead. A typical rotating head (fig. 5.8) seals around the kelly to keep mud, air, or gas from spraying out of the hole through the rotary table. Rotating heads are useful for circulating out low-volume, high-pressure gas kicks while continuing to drill ahead with underweight mud.

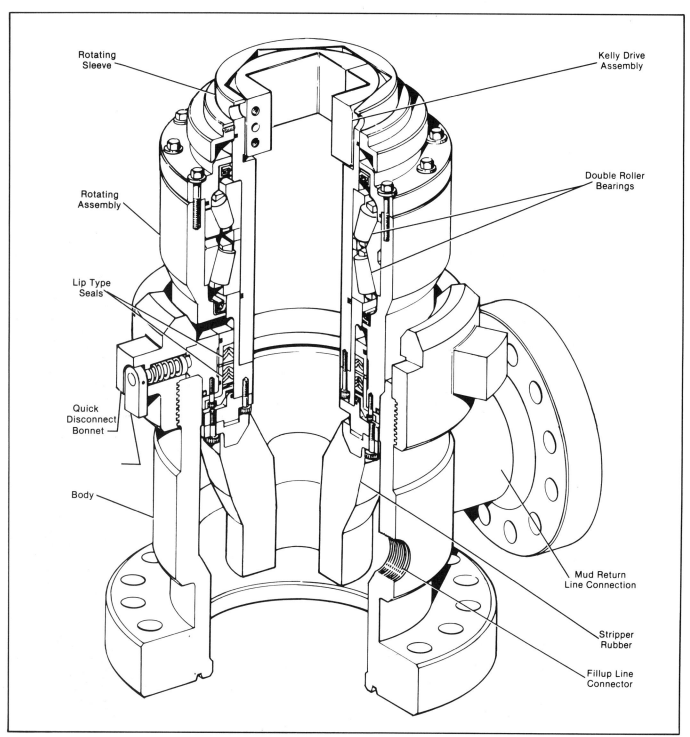

Figure 5.8. Cutaway view of rotating head (*Courtesy of NL Shaffer*)

When pipe is in the hole, it must be possible to close off not only the annulus but also the inside of the drill string; when a kick occurs, an *internal blowout preventer* is usually installed for this purpose (fig. 5.9). Internal preventers open to allow fluid to be pumped down the drill string but close under pressure from below. Other drill string blowout prevention devices are the *kelly cock* (fig. 5.10), which can be used to shut off well pressure if the swivel packing or drilling hose fails, and the lower kelly cock or *drill stem safety valve* (fig. 5.11), which may be closed to permit removing the kelly and installing a circulating line or a back-pressure valve while killing a well.

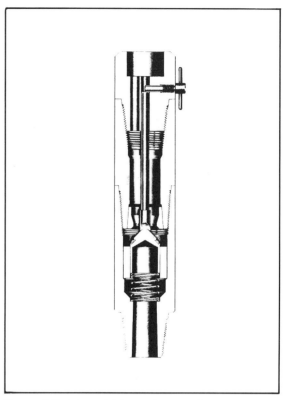

Figure 5.9. Internal blowout preventer

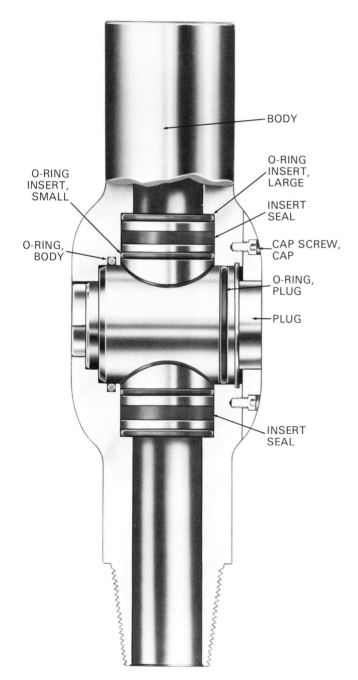

Figure 5.10. Upper kelly cock

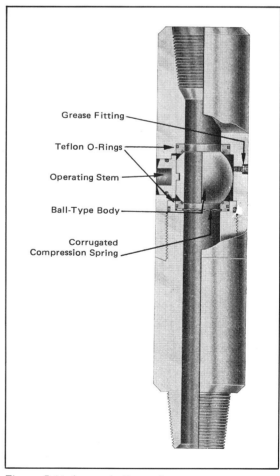

Figure 5.11. Lower kelly cock

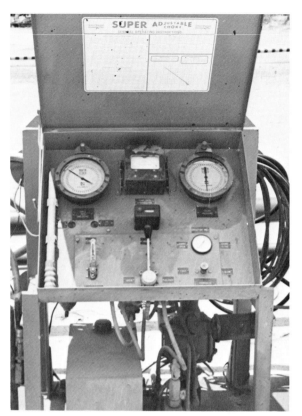

Figure 5.13. High-pressure choke manifold

Figure 5.14. High-pressure (10,000 psi) choke

During a kill, pressure is regulated by holding back-pressure on the casing with a *choke*—a high-pressure adjustable-opening valve through which well fluids flow out of the BOP stack (fig. 5.12). Multiple chokes are essential for positive control; the choke manifold is usually mounted outside the rig substructure (fig. 5.13). Hydraulic chokes can be adjusted from a remote control panel, which also shows casing and drill pipe pressures, pump speed, and cumulative pump strokes (fig. 5.14).

Figure 5.12. Control unit for adjustable choke

Most blowout preventers and many valves and chokes are opened and closed hydraulically. A small pump, powered by air, gasoline, or electricity, pumps hydraulic fluid into tanks known as *accumulators,* where it is stored under pressure (fig. 5.15). Hydraulic lines from the accumulators to the BOP stack are opened and closed from the *master control panel* near the driller's position (fig. 5.16) or from the *remote control panel* of the *pump-accumulator unit,* located a safe distance from the wellbore in case of a blowout (fig. 5.17).

Figure 5.16. Master BOP control panel (*Courtesy of NL Koomey*)

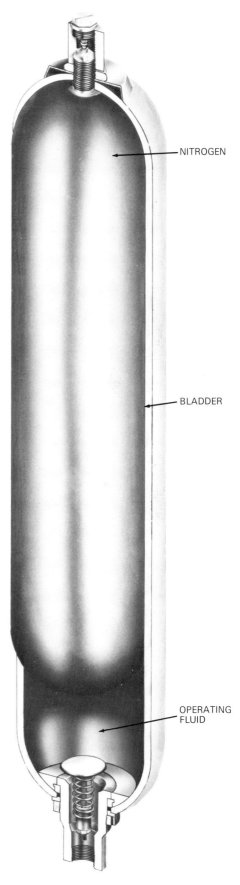

Figure 5.15. Cutaway view of bladder-type accumulator (*Courtesy of NL Koomey*)

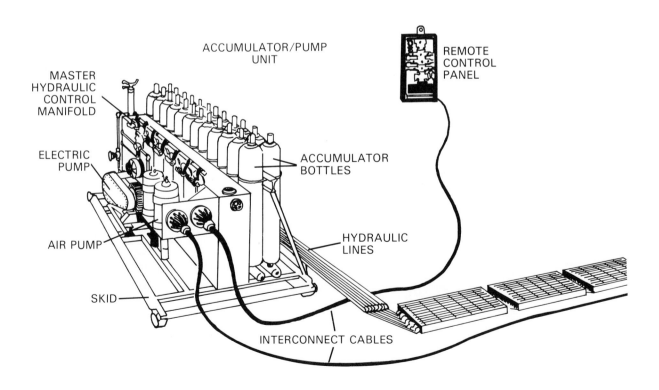

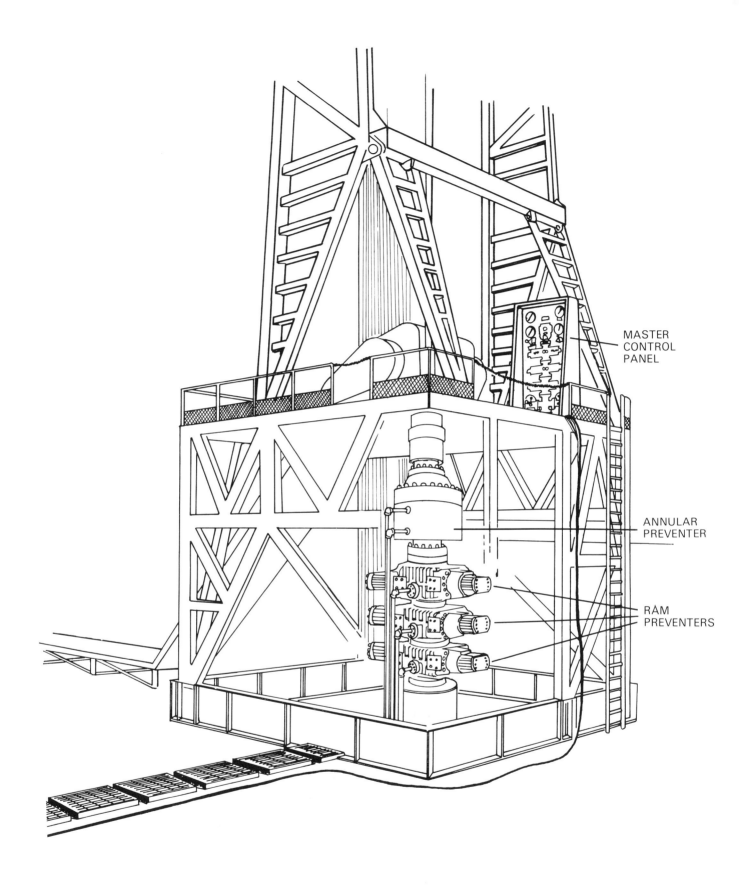

Figure 5.17. Typical arrangement of BOP and control systems (*Courtesy of NL Koomey*)

Driller's Method

Once the well has been closed in and the mud weight needed to kill the kick has been calculated, two actions must be taken: (1) the intruding fluid must be circulated out, and (2) heavier mud must be circulated in. These actions can be taken separately or together. In the *driller's method,* they are done separately.

Drilling fluid is circulated at a constant, reduced pump speed (usually ⅓ to ½ normal speed) and circulating pressure. There are several reasons for this practice:

1. Normal circulating pressure plus the kick pressure might exceed the pressure rating of the pump.
2. Fluid volume and pressure at the normal speed might require more power than available to run the pump.
3. Higher flow rates might cause excessive friction pressure in the annulus, leading to lost returns.
4. Higher flow rates might not give the operator time to adjust the choke as necessary to maintain constant BHP.
5. The crew might not be able to weight up the mud fast enough.

In the driller's method, the kick is first circulated out at the reduced pump speed, using mud of the weight in use when the kick occurred. To prevent further entry of formation fluids, the choke is opened or closed to hold drill pipe pressure constant; opening the choke (increasing its size) lowers pressure, and closing (decreasing its size) raises pressure. The intruded fluid is circulated up the annulus to the wellhead and out of the well.

In the second step, kill-weight mud is circulated in to replace the lighter mud, again at the reduced pump speed. This time, casing pressure is held constant by adjusting the choke while the drill pipe is being filled with the new mud. Once the new mud reaches the bit, the choke is adjusted to keep drill pipe pressure constant. Casing pressure gradually drops as the annulus fills with kill-weight mud.

Circulating out the kick. The first circulation is started by opening the choke and bringing the pump up to the preselected speed, keeping casing pressure constant by adjusting the choke. When the pump is at the required speed and circulating pressure has stabilized, DPP is maintained at (1) the value obtained by adding SIDPP to the preselected reduced circulating pressure (fig. 5.1, part 3), or (2) the value observed when the pump reaches the preselected constant speed while casing pressure is maintained.

Lag time between choke adjustment and pressure changes at the standpipe should be taken into account. This delay is about one second per 1,000 ft of fluid column down the annulus and back up the drill stem, or 2 seconds per 1,000 ft of measured depth with the bit on bottom. In a

10,000-ft well, for example, 20 seconds may elapse after the choke size is reduced before the pressure increase shows up on the drill pipe gauge, depending upon the amount of formation fluid and the weight of the mud in the annulus. Failure to allow for this delay can lead to overcorrecting and excessive pressure fluctuation and result in formation breakdown, additional fluid entry, and an underground blowout.

Pump speed must be kept constant; it can be affected by pressure changes and other factors and so must be carefully monitored during circulation. However, the pump can be stopped at any time and the well shut in without danger to the kill operation. New SIDPP and SICP readings can then be used when circulation is reestablished.

Figure 5.18 illustrates what happens as the freshwater kick shown in figure 4.21 (previous lesson) is circulated out of the well. (For purposes of illustration, this kick is far larger than should ever occur in an actual drilling situation. To simplify the example and show hydrostatic pressures only, the well is shown shut in, which is not a normal condition under standard operating procedures.) Note that as the freshwater intrusion is pumped up the annulus, total hydrostatic pressure at the bottom of the annulus does not change; hydrostatic pressure at any point is the sum of the hydrostatic pressures of all fluids above that point. The intruded water column occupies the same volume at the top of the hole as it did at the bottom because, like most liquids, it is not significantly compressible. Therefore, there is as much mud in the hole as before the kick was circulated to the top.

As soon as the water begins to leave the annulus, it is replaced by 10-ppg mud (fig. 5.19). SICP drops gradually as the new mud exerts more and more hydrostatic pressure against the formation pressure.

After the freshwater kick has been circulated out of the well (fig. 5.20), both drill pipe and annulus are again full of 10-ppg mud, so the hydrostatic pressure at the bottom of each is 5,200 psi. If the well is shut in, formation pressure imposes an additional 300 psi on each fluid column. SIDPP and SICP both read 300 psi—the amount by which the hydrostatic pressures of the fluids in both the drill stem and the annulus underbalance formation pressure.

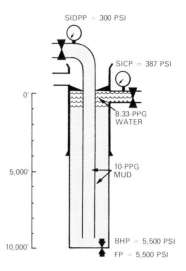

Figure 5.18. Freshwater intrusion being circulated out of well

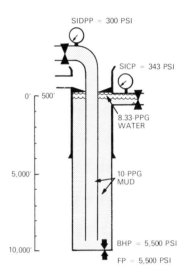

Figure 5.19. Water leaving annulus

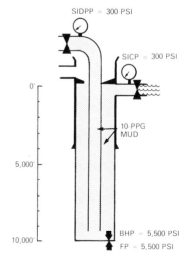

Figure 5.20. Freshwater intrusion circulated out of well

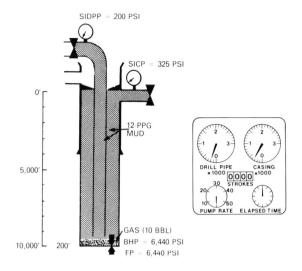

Figure 5.21. Gas kick, shut in

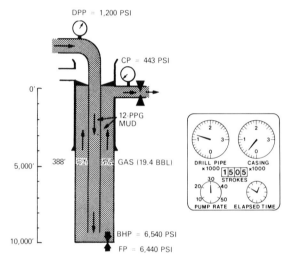

Figure 5.22. Circulating out a gas kick

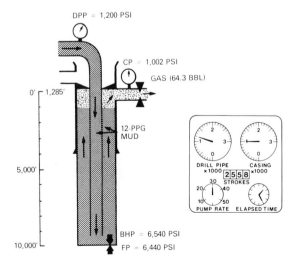

Figure 5.23. Gas kick at surface

Because of its compressibility, gas being circulated out of a well affects well pressures differently. In our hypothetical 10,000-ft well, the hydrostatic pressure of 12-ppg mud in the drill stem (6,240 psi) underbalances formation pressure (6,440 psi) by 200 psi, shown as SIDPP (fig. 5.21).

Pit gain is 10 bbl; assuming an average annular volume of 0.05 bbl/ft, the 10-bbl kick has a vertical length of 200 ft (10/0.05). The 9,800 ft of mud in the annulus exerts a hydrostatic pressure of 6,115 psi. (The hydrostatic pressure contribution of the intruded gas is negligible.) Shutting in the well adds just enough pressure to the annulus to make up the difference between formation and mud hydrostatic pressure—325 psi. The additional pressure shows up on the casing pressure gauge as SICP.

Figure 5.22 shows the kick being circulated out, using a kill-rate pressure of 1,000 psi. Kill-rate pressure plus 200 psi SIDPP equals 1,200 psi, called *circulating drill pipe pressure* or *initial circulating pressure* (step 3 on the worksheet). Pressure loss in the drill stem and bit nozzles is 900 psi, leaving 100-psi circulating pressure to overcome annular fluid friction. BHP is 6,540 psi; but the hydrostatic pressure of the 5,000 ft of mud below the kick is only 3,120 psi. The remainder of the BHP comes from gas pressure in the kick—3,370 psi—and annular friction from 10,000 ft to 5,000 ft (50 psi). Adjusting the choke to maintain standpipe pressure at 1,200 psi allows the gas bubble to expand enough to maintain a constant BHP, overbalancing formation pressure by 100 psi and preventing further intrusion of formation gas.

Although standpipe (drill pipe) pressure is held constant, casing (annular) pressure increases as the gas is circulated toward the surface. At 3,370 psi, the gas occupies a volume of 19.4 bbl (Boyle's law). Its vertical length is now 388 ft, leaving 4,612 ft of mud in the upper annulus. This mud exerts a hydrostatic pressure of 2,878 psi; but the gas is opposing it with a pressure of 3,370 psi. This net upward pressure of 492 psi is opposed by annular friction (50 psi) and by the choke, which holds casing pressure at 442 psi and keeps the well from unloading.

Maximum surface casing pressure and pit gain occur when the gas reaches the top of the annulus (fig. 5.23). By this time the kick has grown to 1,285 ft because it is under less pressure—1,002 psi. The 8,715 ft of mud below the kick exerts 5,438 psi of the total BHP of 6,540 psi; annular friction contributes another 100 psi. The pressure of the gas can now be read directly from the casing pressure gauge.

In figure 5.24, the gas has been circulated out of the well. Shutting in the well now shows that the kick has been eliminated: SICP = SIDPP = 200 psi. The hydrostatic pressure of 12-ppg mud still underbalances formation pressure by 200 psi. (Since the well is no longer being circulated, the 100-psi overbalance from annular fluid friction is gone.) Heavier mud must be circulated in before drilling is resumed.

As gas is circulated up and out of the annulus, there is a pit gain equal to the volume of the gas that entered the borehole plus its expansion as it rises in the hole (fig. 5.25). When all the gas has been circulated out of the well, the mud volume in the pits has returned to its original level. The well is again full of mud.

If the intruding fluid is oil or salt water, pit gain stays constant as the well is circulated. When the formation water or oil has left the casing, clean mud appears at the outlet. The pump can be stopped and the well closed in; both SIDPP and SICP will equal the difference between mud hydrostatic pressure and formation pressure. The well can safely be left shut in as long as necessary while the mud is being weighted up.

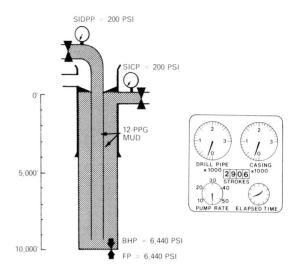

Figure 5.24. Shut-in pressures with gas out of well

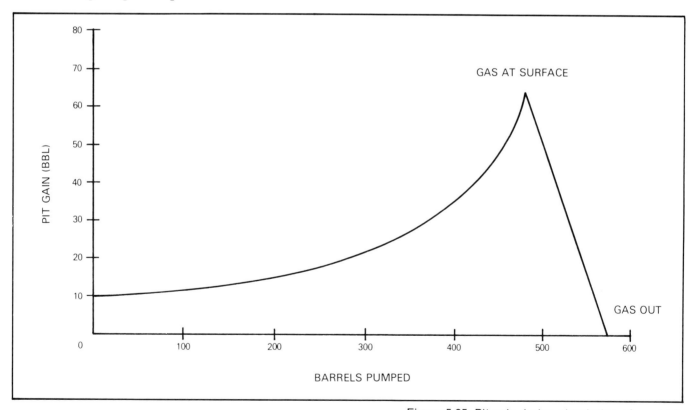

Figure 5.25. Pit gain during circulation of gas kick

173

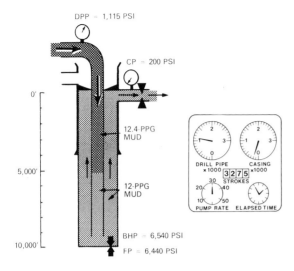

Figure 5.26. Kill-weight mud filling drill stem

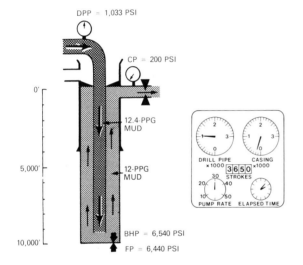

Figure 5.27. Kill-weight mud arriving at bit

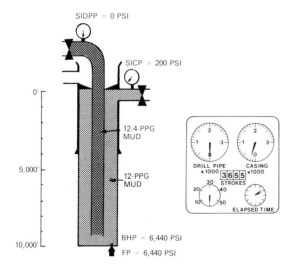

Figure 5.28. Drill stem filled with kill-weight mud, well shut in

Circulating in new mud. The new mud weight required to control formation pressure is calculated by using the formula shown in part 4 of the worksheet:

$$MWI = \frac{20 \times SIDPP}{TVD}$$

$$= \frac{20 \times 200}{10,000}$$

$$= \frac{4,000}{10,000}$$

$$= 0.4 \text{ ppg.}$$

Mud weight must be increased by 0.4 ppg to balance the 6,440-psi formation pressure that caused the kick. The new mud weight, 12.4 ppg, is recorded in part 5.

(The above formula produces about 4% greater mud weight than needed to balance formation pressure when the pump is stopped. A more precise formula—

$$MWI = \frac{19.23 \times SIDPP}{TVD}$$

can be used. However, it may be desirable to use the former to add a safety factor, especially before making a trip.)

Kill-weight mud can be circulated in at any constant pump speed, depending upon how fast it can be mixed in the pits. Normally, however, the preselected pump rate is used. While the heavier fluid is being pumped down to the bit, casing pressure is held constant at the last observed SICP (fig. 5.26). This procedure holds sufficient backpressure to control formation pressure without risking formation breakdown.

Normally, the pressure required to circulate at a given rate increases with mud weight. During a kill, however, pressure at the standpipe drops as new mud fills the drill stem (figs. 5.26 and 5.27) because the heavier mud in the drill stem helps the pump push the lighter mud out of the annulus. When kill-weight mud has reached the bit, the hydrostatic pressure of the mud in the drill stem balances formation pressure, and the drill pipe pressure reading reflects only circulating pressure. Therefore, even though the pump is circulating heavier mud, final circulating pressure (FCP) will be less than the initial circulating pressure, but greater than kill-rate pressure (KRP). This relationship is shown by the formula in part 6 of the well control worksheet:

$$FCP = KRP \times \frac{NMW}{OMW}$$

where
- FCP = final circulating pressure;
- KRP = kill-rate pressure;
- NMW = new mud weight; and
- OMW = original mud weight.

The new mud weight required to balance formation pressure is 12.4 ppg, so the final circulating pressure will be—

$$FCP = 1{,}000 \times \frac{12.4}{12.0}$$
$$= 1{,}000 \times 1.033$$
$$= 1{,}033 \text{ psi}.$$

In figure 5.28, with the well shut in again (not standard operating procedure in a kill), the hydrostatic pressure of the mud in the drill stem is now 6,440 psi. This pressure balances formation pressure, so SIDPP = 0 psi. The annulus, however, still full of 12-ppg mud, needs an additional 200 psi (SICP) to balance formation pressure. (Compare this situation with the one shown in figure 5.24.)

As pumping resumes, the choke is manipulated to keep drill pipe pressure at 1,033 psi while the annulus fills with new mud (fig. 5.29). With kill-weight mud displacing the lighter mud, casing pressure drops steadily.

When the annulus has filled with kill-weight mud (fig. 5.30), casing pressure has dropped to zero. The 1,033-psi drill pipe pressure (FCP) is lost to friction down the drill stem, through the bit jets, and up the annulus, leaving no pressure on the casing at the surface.

Circulation can now be halted; if the mud has been properly weighted, the well will not flow. With the shut-in well full of 12.4-ppg mud, both SIDPP and SICP are reduced to zero (fig. 5.31). Hydrostatic pressure alone now keeps formation pressure under control. If pressure remains, however, mud weight can be recalculated, mud density increased, and the circulation repeated.

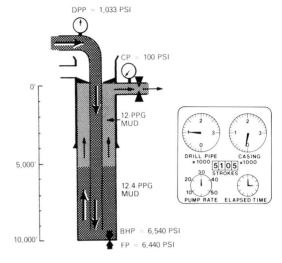

Figure 5.29. Annulus filling with kill-weight mud

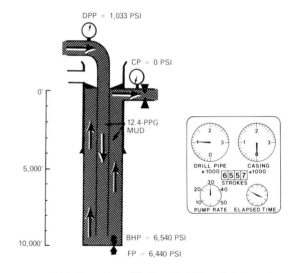

Figure 5.30. Annulus filled with kill-weight mud

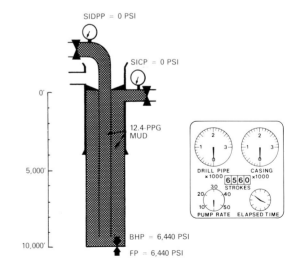

Figure 5.31. Well shut in with kill-weight mud

Figure 5.32 shows how pressures change as the gas kick in figure 5.21 is killed by using the driller's method. Casing pressure must not exceed *maximum allowable annular surface pressure* (MAASP), usually considered 80% of casing design burst strength. Keeping casing pressure below MAASP is the best reason for detecting a kick early; the smaller the kick, the lower the maximum casing pressure will be during the kill.

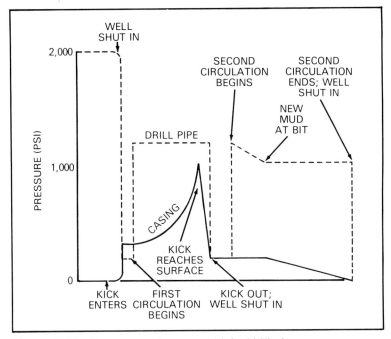

Figure 5.32. Pressures during a kick (driller's method)

To summarize: In the driller's method, casing pressure is held constant while the drill pipe is filled with kill-weight mud. Once this mud reaches the bit, attention is shifted to drill pipe pressure, which is then held constant until all original-weight mud is out of the annulus.

If a separate pit and mixing pump are available, mud weight can be increased while the kick is being circulated out with the original mud. Kill-weight mud can then be started down the drill pipe as soon as the kick has left the annulus, saving time and reducing the chances of another kick.

Part 7 of the worksheet, the calculation of surface-to-bit strokes, shows when kill-weight mud can be expected to reach the bit. Drill pipe pressure is kept constant while the kick is circulated out of the annulus; casing pressure is held constant while the drill stem is filled with weighted-up mud. When kill-weight mud arrives at the bit, drill pipe pressure will have dropped to the calculated FCP (part 6), which is maintained by choke adjustment until new mud fills the annulus. It is often useful, though not essential, to complete part 8, which shows how DPP will change as the kill-weight mud fills the drill stem.

Often, the reason the driller's method is used is that it requires fewer calculations than other methods. It can be started as soon as the well is shut in, allowing the kick to be circulated out without having to wait for mud to be weighted up. It also reduces the hazard of uphole gas migration with the well shut in. During a gas kick, however, the driller's method causes higher surface and casing-seat pressures than other methods because kill-weight mud is not in the annulus to help offset formation pressure, and the difference must be compensated for by holding extra back-pressure on the choke while the gas is circulated out.

Wait-and-Weight Method

The most common alternative to the driller's method, and in some areas the standard well control procedure, is the *wait-and-weight* method, so named because rig personnel wait until the mud has been weighted up before beginning to circulate the kick out of the well. If kill-weight mud can be prepared without delay, and if uphole gas migration in the meantime does not cause pressures to rise excessively, the well can be kept shut in until the mud is ready.

The wait-and-weight method combines the two circulations of the driller's method into one. The required kill-weight mud is mixed before circulation is started. Part 8 of the worksheet must be completed (fig. 5.33). Since weighted mud is introduced into the drill stem at the beginning of the process, both drill pipe pressure (DPP) and casing pressure (CP) change continuously until new mud reaches the bit; the choke operator must know how much DPP to hold with the choke at any moment. In the example, DPP will be 1,133 after 300 strokes (10 min) and 1,067 after 600 strokes (20 min). At 750 strokes (25 min), weighted mud will reach the bit and DPP will be at FCP: 1,033 psi.

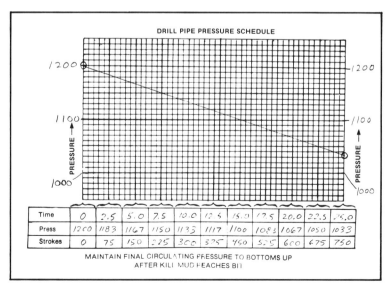

Figure 5.33. Well-control worksheet, part 8 (completed for wait-and-weight method)

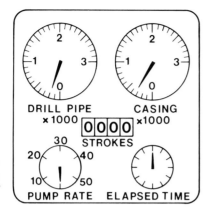

Figure 5.34. Gas kick, shut-in pressures

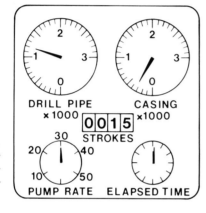

Figure 5.35. Gas kick, initial circulating pressure

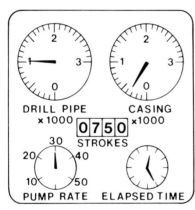

Figure 5.36. Gas kick, kill-weight mud at bit (wait-and-weight method)

If for some reason the well is circulated at a higher or lower pump speed than that recorded in part 1, the recorded SICP is held with the choke while DPP is read. The observed DPP is then considered to be ICP (initial circulating pressure) and is either maintained on the standpipe gauge until the kick is out of the well (driller's method) or adjusted gradually downward to FCP as kill-weight mud approaches the bit (wait-and-weight method).

Figures 5.34–5.40 demonstrate pressures and gauge readings that might be observed while the gas kick shown in figure 5.21 is controlled by the wait-and-weight method. With the well shut in, DPP = 200 psi and CP = 325 psi (fig. 5.34). Once sufficient mud has been weighted up in the pits to fill the hole and control formation pressure, the pump is started up to the predetermined kill rate (30 strokes/min). The choke is adjusted to keep casing pressure constant, momentarily, at 325 psi; when pump speed and pressure have stabilized, the choke is manipulated to keep ICP (1,200 psi) on the drill pipe pressure gauge (fig. 5.35).

As kill-weight mud is pumped down the drill string, ICP is gradually allowed to fall to FCP. When the pump-stroke count has reached 750, kill-weight mud has reached the bit (fig. 5.36) and 1,033 psi should be held on the drill pipe.

FCP is maintained as circulation continues (fig. 5.37). Maximum casing pressure occurs when the kick reaches the surface (fig. 5.38); this event may be accompanied by a great deal of noise and vibration as the expanding gas escapes through the choke. Once the kick is out, casing pressure falls dramatically (fig. 5.39); then it drops slowly but steadily until new mud reaches the choke, 750 strokes after the kick is out (fig. 5.40).

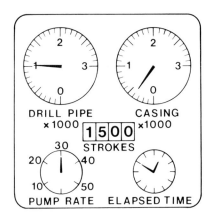

Figure 5.37. Gas kick, annulus partly filled with kill-weight mud (wait-and-weight method)

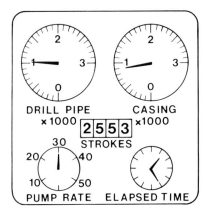

Figure 5.38. Maximum casing pressure (wait-and-weight method)

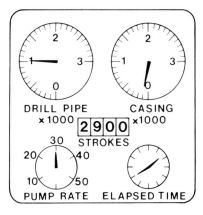

Figure 5.39. Gas kick out of annulus (wait-and-weight method)

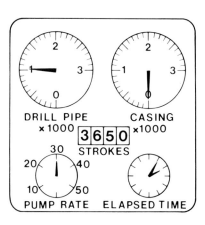

Figure 5.40. Well filled with kill-weight mud

The wait-and-weight method, like the driller's method, maintains sufficient BHP to keep additional formation fluids from entering the hole. However, it may allow lowered casing-seat and casinghead pressures. Figure 5.41 shows why. (To simplify calculations, the well is shown shut in at each stage. Normally, the well would be circulated continuously during a kick control operation. There is also assumed to be no gas-cutting of the drilling mud, and the hydrostatic pressure of the gas is ignored.) The annulus holds more fluid than the drill stem, so when the drill stem has been filled with kill-weight mud (G), only part of the old mud has left the annulus and the kick has been circulated only partway up the hole. Maximum pressure at the casing seat occurs when the gas bubble reaches it. At this stage in the wait-and-weight method (H), the annulus below has partly filled with kill-weight mud. This mud exerts more hydrostatic pressure on the bottom than the same amount of the original mud; therefore less casinghead pressure is needed to hold the required BHP. The gas and the casing seat are under less pressure at this point than in the driller's method (C), where more of the BHP must be maintained by adjusting the choke.

Casinghead pressure also peaks with the arrival of the kick; but maximum CP is less with the wait-and-weight method (J) than with the driller's method (E). The lower pressure allows a greater safety margin below casing burst pressure.

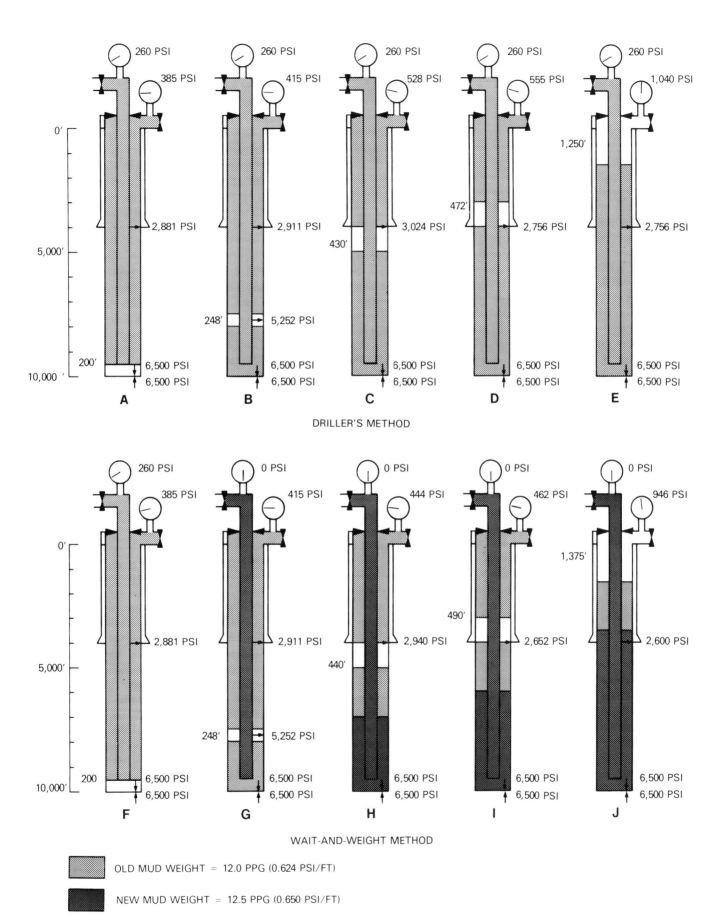

Figure 5.41. Well pressure distributions with a gas kick under two methods of kick control

Concurrent Method

The *concurrent method* (also called the *circulate-and-weight method*) is the fastest method of controlling a kick. Immediately after stabilized SIDPP and SICP are recorded, circulation is begun at the reduced rate. Mud is weighted by stages and pumped in as soon as each batch is ready. Maximum casing pressure is usually less than in the driller's method but greater than in the wait-and-weight method. More calculations and closer supervision of rig personnel are required than in either of the other methods.

Part 8 of the worksheet is filled out differently in the concurrent method (fig. 5.42). As in the wait-and-weight method, ICP and FCP are connected by a straight line; instead of time, however, intermediate mud weight values are entered along the bottom of the graph. As soon as the derrickman finishes mixing a batch of mud, he notifies his supervisor that mud of a particular weight is ready to be pumped down the drill pipe. Beneath that mud weight on the graph, the supervisor records the number of pump strokes since circulation was started (second line), adds the number of surface-to-bit pump strokes (previously calculated), and records the total on the third line. When the pump stroke count reaches the number shown on the third line, indicating that the new mud is at the bit, the drill pipe

Figure 5.42. Pressure graph (concurrent method)

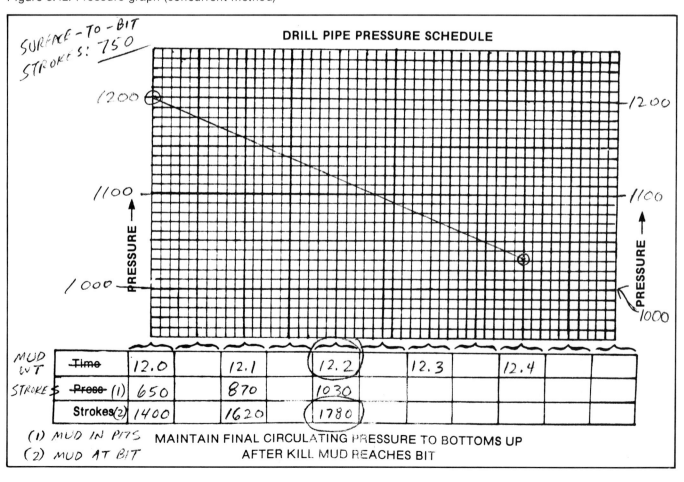

(circulating) pressure should be maintained as shown on the graph directly above.

As indicated in figure 5.42, when the derrickman calls out a mud weight of 12.2 ppg, the supervisor records the pump-stroke count: 1,030. He adds 750 strokes (to fill the drill stem) and records 1,780 below the 1,030. As the pump stroke count reaches 1,780, he adjusts the choke to hold 1,110 psi on the drill pipe. Finally, when 12.4-ppg kill-weight mud reaches the bit (750 strokes after the derrickman has it ready), circulating drill pipe pressure should read 1,033 psi (FCP).

Incorrect Kick Control Methods

Pulling into the casing. When a well begins to flow, a common mistake is to pull the drill pipe all the way up into the casing to keep it from becoming stuck. But time wasted pulling up into the casing allows additional fluid to enter the wellbore. When handling a kick, the best way to prevent sticking is to minimize formation fluid entry. Standard practice should be to raise the kelly just enough to clear the first tool joint above the rotary table. If the pipe must be pulled up more than about 100 ft, the well-killing problem becomes more complicated, requiring heavier mud than that required if the drill pipe remains near bottom.

Constant pit-level method. The constant pit-level method is practical only for gas-free kicks. If gas is present, pit level can be kept constant only by not allowing the gas to expand as it comes uphole. As shown in figure 4.21 (previous lesson), casing and bottomhole pressures quickly become excessive under these conditions, leading to formation breakdown, lost circulation, an underground blowout, and complete loss of the well. The primary means of control should always be to maintain constant bottomhole pressure while pumping at a constant rate.

Constant choke-pressure method. Like the constant pit-level method, the constant choke-pressure technique has some basis in fact. For a saltwater flow with little or no gas, the kick can be circulated out while constant backpressure is maintained at the wellhead. Choke pressure must be reduced during the second circulation, since the annulus is filling with heavier mud. However, when gas is circulated out with constant choke pressure, it expands more than it should as it rises in the annulus, leaving too little mud in the hole. BHP becomes insufficient to control formation pressure, allowing the well to kick again.

Excessive mud weight. If the drilling fluid used has a weight much greater than that needed to control formation pressure (based on SIDPP), the wait-and-weight method may produce excessive pressures at bottom and especially at the casing seat, which may fail. The procedure can be modified to compensate for this effect, but nothing is gained by "overkilling" the well.

PREVENTER DRILLS

BOP drills accomplish several things:

1. They increase crew alertness to signs of an impending kick.
2. They decrease reaction time and keep kick size small.
3. They familiarize the crew with BOP equipment operation.
4. They develop crew coordination.
5. They instill confidence and forestall panic.
6. They test the operational readiness of BOP equipment.

The driller is the key to successful well control. He is responsible for directing the crew's actions when a kick is first detected. Often he is the first to know, because his instruments may provide the first indication of a kick—increased penetration rate, decreased circulating pressure, or pit gain.

Once the crew has been instructed and the equipment demonstrated, drills should be initiated without advance notice to the driller or his crew. The toolpusher can start the drill by raising (or lowering) the mud-pit float, causing an indication on the driller's pit-level indicator. The toolpusher notes the time. When the driller gives notice that he has seen the pit-level change, the toolpusher records the driller's reaction time. Finally, he records how quickly and skillfully the driller and his crew close in the well. The time it takes to (1) become aware of the kick and (2) shut in the well should be as short as possible. Thirty seconds wasted can allow the kick to grow by many barrels, causing increased shut-in pressures and greater danger while the kick is being circulated out of the well.

Equally important is the ability of the driller and his crew to respond to different situations. Most operators conduct drills under at least two conditions: (1) drilling, and (2) making a trip. Some companies also conduct drills with collars in the preventers and with no pipe in the hole.

If pit-level instruments are not available, the toolpusher can simply tell the driller to start a drill. However, during a real kick, pit gain usually develops gradually and silently. The alertness of the driller and crew to subtle clues is the key to effective well control.

LESSON 5 QUESTIONS

Put the letter for the best answer in the blank before each question.

_____ 1. The first action the driller takes when he suspects a kick is to—
   A. close the BOP.
   B. read and record the stabilized SIDPP.
   C. shut down the pump.
   D. raise the kelly.

_____ 2. The main difference between shutting in the well with the bit on bottom and shutting in the well during a trip is—
   A. installing an inside BOP in the drill string.
   B. removing all pipe from the hole immediately.
   C. using reverse circulation.
   D. all of the above actions.

_____ 3. In shutting in a well, a *soft close* means—
   A. closing the BOP gradually.
   B. closing only the annular preventer.
   C. closing the annular preventer with moderate pressure.
   D. closing the BOP with the choke line open.

_____ 4. In a shut-in well, gas rising in the annulus causes—
   A. SICP to increase and SIDPP to decrease.
   B. SICP to decrease and SIDPP to increase.
   C. SICP and SIDPP to increase.
   D. SICP and SIDPP to decrease.

_____ 5. An annular preventer *cannot*—
   A. seal off open hole.
   B. cut through drill pipe.
   C. close off around the kelly.
   D. seal off around rotating pipe.

_____ 6. Pipe rams—
   A. cut through drill pipe.
   B. seal off open hole.
   C. seal around the kelly.
   D. seal off the annulus around drill pipe.

_____ 7. Blind rams—
   A. seal off the annulus around any size or shape of drill pipe.
   B. seal off open hole.
   C. seal off the inside of the drill string.
   D. seal around the kelly.

_____ 8. A blowout preventer designed for routine drilling under pressure is –
   A. the annular preventer.
   B. the blind ram.
   C. the rotating head.
   D. the internal blowout preventer.

_____ 9. Which of the following devices does *not* control flow up the inside of the drill string?
   A. Internal blowout preventer
   B. Kelly cock
   C. Blind ram
   D. Drill stem safety valve

_____ 10. BOP operating and control equipment is located –
   A. next to the mud pits and mud pumps.
   B. beneath the derrick floor.
   C. a safe distance from the wellbore.
   D. in the doghouse.

_____ 11. When the choke is adjusted while a 6,000-ft well is circulated under pressure, how long will it take for the pressure change to show up on the drill pipe gauge?
   A. 6 seconds
   B. 12 seconds
   C. 24 seconds
   D. 6 minutes

_____ 12. A freshwater intrusion will expand at least 50 percent as it rises from the bottom of a 10,000-ft well. (T/F)

_____ 13. During the first circulation of the driller's method –
   A. drill pipe pressure gradually falls.
   B. drill pipe pressure gradually rises.
   C. pump speed is gradually reduced to keep pressure constant.
   D. the choke is adjusted to keep drill pipe pressure constant.

_____ 14. As kill-weight mud is pumped down the drill string during the second circulation of the driller's method –
   A. drill pipe pressure gradually falls.
   B. drill pipe pressure gradually rises.
   C. drill pipe pressure does not change.
   D. drill pipe pressure rises, then falls to zero.

(*Questions 15–22*) In the shut-in 10,000-ft well diagrammed at right, what is the pressure –

_____ 15. at point 15?
_____ 16. at point 16?
_____ 17. at point 17?
_____ 18. at point 18?
_____ 19. at point 19?
_____ 20. at point 20?

A. 3,210 psi
B. 4,354 psi
C. 6,070 psi
D. 208 psi
E. 350 psi
F. 3,418 psi
G. 558 psi
H. 1,494 psi

_____ 21. What is the formation pressure at 10,000 ft?

_____ 22. What will be the SICP when the saltwater kick has been circulated out?

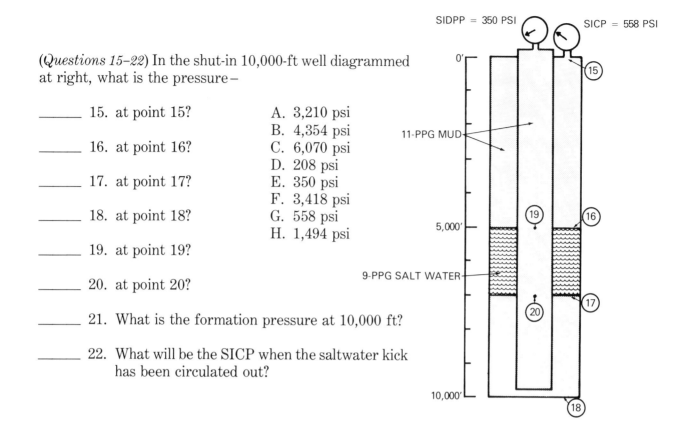

(*Questions 23–25*) In using the driller's method, a kick has been circulated out of a 17,500-ft well, and the well has been shut in. SIDPP and SICP are both 475 psi. Mud weight is 13.6 ppg. Standpipe pressure during circulation at 30 strokes/min was 1,575 psi.

_____ 23. What is the formation pressure?
  A. 12,851 psi
  B. 12,376 psi
  C. 9,100 psi
  D. 7,072 psi

_____ 24. What is the mud weight required to kill the well?
  A. 10.5 ppg
  B. 13.1 ppg
  C. 13.6 ppg
  D. 14.1 ppg

_____ 25. What will be the new circulating pressure (FCP) after kill-weight mud reaches the bit?
  A. 1,632 psi
  B. 1,496 psi
  C. 1,140 psi
  D. 925 psi

_____ 26. Why does the wait-and-weight method allow a lower maximum casinghead pressure than the driller's method?
   A. It uses heavier kill-weight mud.
   B. Bottomhole pressure is lower with the wait-and-weight method.
   C. Kill-weight mud is present in the annulus at the same time as the kick.
   D. Pressure in the well drops while kill-weight mud is being mixed.

_____ 27. What is the main advantage of the concurrent method?
   A. It is faster than other methods.
   B. It produces the lowest casinghead and casing-seat pressures.
   C. It allows use of lighter kill-weight mud.
   D. All of the above statements are true.

_____ 28. In figure 5.42, if the derrickman calls out "12.3" at 1,250 strokes, what drill pipe pressure should be held when the pump-stroke counter reads 2,000?
   A. 1,100 psi
   B. 1,075 psi
   C. 1,033 psi
   D. 750 psi

_____ 29. Most kicks can be handled by using the constant pit-level method. (T/F)

To answer questions 30–40, use the instrument readings shown at right and the information shown on the worksheets in figures 5.1 and 5.33. Unless a set of answers is provided, choose your answer for each question from the following set:

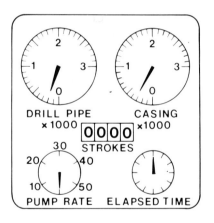

   A. increase choke size.
   B. decrease choke size.
   C. increase pump rate.
   D. decrease pump rate.
   E. continue; everything is OK.
   F. stop the pump and shut in the well.

Imagine that you have to control a well that has taken a kick and been shut in. You have recorded the SIDPP and SICP shown on the console (right). You decide to use the wait-and-weight method.

_____ 30. When kill-weight mud is ready, you start the pump. To bring drill pipe pressure to the initial circulating pressure (ICP), you should –

_____ 31. Check your gauge readings with the graph. You should –

_____ 32. Everything appears to be going normally, so you turn the operation over to an inexperienced floor man for a few minutes to have a cup of coffee in the doghouse. When you return, the situation is as shown. You should –

_____ 33. You check your worksheet again and everything seems to be going well; then you notice this situation. You should –

_____ 34. Immediately after this action, you will probably have to –

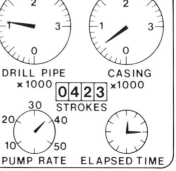

189

_____ 35. Your drill pipe pressure has dropped to 1,033 psi. You should—

_____ 36. What has happened?
   A. The formation has broken down.
   B. A blowout is in progress.
   C. Kill-weight mud has reached the bit.
   D. You are not holding enough pressure with the choke.

_____ 37. Casing pressure has stayed constant, but drill pipe pressure has changed. You should—

_____ 38. Drill pipe pressure is back to normal, but casing pressure is rising rapidly and the rig is beginning to shake. You should—

_____ 39. What is happening?
   A. The well is blowing out.
   B. The kick has reached the surface.
   C. The pump is malfunctioning.
   D. The drill pipe has ruptured.

_____ 40. Casing pressure has dropped; pit level has returned to normal. You should now—
   A. resume drilling.
   B. stop the pump, recalculate the mud weight, and circulate again.
   C. stop the pump and see if mud flows from the well.
   D. close the choke to increase drill pipe pressure to 1,200 psi.

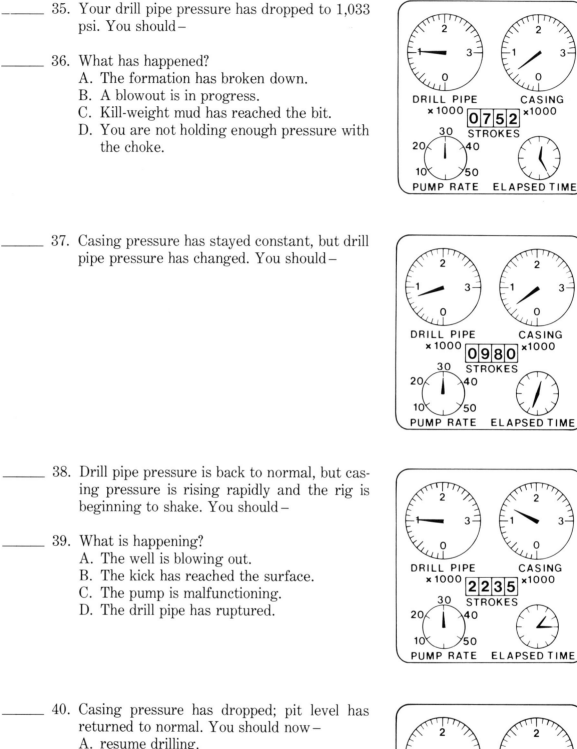

Lesson 6
Optimization

Introduction

Bits

Weight on Bit and Rotary Speed

Drilling Fluids

Bit Hydraulics

Formation Properties

Computerized Optimization

Lesson 6
OPTIMIZATION

INTRODUCTION

Veteran drillers will tell you there is just one rule in drilling: set the bit on bottom and turn it to the right. No doubt that is the main thing, but the driller has to know much more before he starts drilling. He must consider weight on bit, rotary speed, drilling fluid properties, and formation characteristics, to name just a few things. He must continue to evaluate these factors during the actual drilling, just as they are weighed while planning the well.

It is not often that everything can be ideal in drilling. Technical decisions involve many compromises that depend on the drilling conditions. But whatever the conditions, every drilling contractor has the same purpose in mind: to drill a usable well to the operator's specifications for the *lowest possible cost*. In drilling, time is money, so effective least-cost drilling means drilling the most usable hole possible to the pay zone in the shortest time. Selecting and using the best combinations of equipment and techniques to accomplish this goal is called *optimization*.

Optimization begins with good well planning. Whether the well is a wildcat or an inside well in a developing field, it can be assumed that the contractor has been studying the site's potential for quite a while. He will generally provide the driller with the information he has on hand. However, if a driller wants more data on a proposed well, there are other sources to look to. Information from other wells in the area may include geologic maps, bit records, electric logs, daily drilling reports, and mud recaps. Before he spuds in, the driller should take the time to learn as much as possible about drilling costs in nearby wells. Rig costs and bit costs may vary, but in every case, correct drilling practices can keep the rate of penetration high and the total well costs down.

A good driller is concerned with six basic factors that affect the rate of penetration (ROP):

1. bit type;
2. weight on bit;
3. rotary speed;
4. drilling fluid properties;
5. bit hydraulics; and
6. formation properties.

The last of these factors, formation properties, is the one that the driller is not able to control. The other five can be adjusted to achieve good penetration rates. The amount of variation possible depends upon rig and equipment capabilities. For example, a rig's capacity for hoisting and handling drill collar weight, turning the rotary, and pumping the drilling fluid determines the weight that can be imposed on the bit, the speed at which the drill string can be rotated, and the effectiveness of the bit hydraulics. The optimized drilling program is the one that most efficiently combines the controllable drilling factors to overcome the formation properties, thereby ensuring that a usable hole is drilled effectively.

BITS

Factors in Bit Selection

Some drillers feel that bit performance is what making hole is all about. Indeed, the bit is the critical part of drilling as long as it is an effective cutting tool in the formations being penetrated. Sometimes the right bit is in the hole at the wrong time. For example, a long-tooth bit that is just right for soft formations is all wrong for hard lime streaks or sandstone stringers that may be encountered unexpectedly. Or a milled-tooth bit that cuts shale faster than a carbide-insert bit can be wrecked by an unexpected streak of chert.

Formation data, mud and casing programs, hole size, and other variables will also influence the choice of bit size and type. Bit records, mud reports, drilling records, and trouble reports from control wells in the area are often used in selecting the proper bit and corresponding operating procedures. Bit records are particularly useful; these forms show pertinent facts of well location, rig capability, mud type, and (most important) vital statistics of each bit run in the course of drilling the hole, including bit type, jet nozzle sizes, footage drilled, dull-bit condition, and hours run.

Bit Types

A bit should be run in the kind of formation best suited to its construction and in a manner that will consistently obtain the best results. However, if much variation in formation characteristics is expected, a bit that can handle different rocks moderately well may save time by reducing the number of trips needed to change bits.

Bit design falls into one of three general categories:

1. roller cone, or rock, bits;
2. diamond bits; and
3. drag bits.

Rock bits have two to four cones with several rows of steel (milled) teeth or tungsten carbide inserts. The number, placement, and composition of the teeth or inserts, as well as the angle of cones, depend upon the type of formation for which the bit is designed (Table 6.1).

Soft-formation bits have long teeth that are protected against abrasion by deposits of tungsten carbide. This protection works well in softer rocks, but running the bit in even moderately hard rocks can chip off hard metal and break teeth. Soft-formation bits are also built with cones offset to scrape the formation; a hard formation quickly dulls any bit that has much offset.

Dense, hard formation must be drilled by crushing. Bits that make hole by the crushing action of the bit teeth must be used at slow rotary speed to allow time for the bit to penetrate the material and to prevent serious bit damage from shock loads. Conversely, bits that make hole by shearing or gouging the formation, such as some drag bits and soft-formation roller cone bits, can be run at high rotary speed without damage.

Insert bits are durable and versatile. They can cost two or three times as much as milled-tooth bits, but because they last longer they can provide more consistent rates of penetration (especially in abrasive streaks) and cut down on the number of trips. A tungsten carbide insert bit, for example, can drill out a cement plug and float collar and continue efficiently into the formation. Table 6.2 shows different combinations of features for various formation types.

TABLE 6.1

MILLED-TOOTH BIT SPECIFICATIONS RELATIVE TO FORMATION TYPE

Formation Type	Tooth Specifications	Degrees Offset
Very soft	Hardfaced tip	3-4
Soft	Hardfaced side	3-4
Medium	Hardfaced side	1-2
Medium-hard	Case hardened	1-2
Hard	Case hardened	0
Very hard	Case hardened, circumferential	0

SOURCE: Jack C. Estes, "Selecting the Proper Rotary Rock Bit," *Journal of Petroleum Technology,* November 1971 (Copyright SPE-AIME 1971.)

TABLE 6.2

INSERT BIT SPECIFICATIONS RELATIVE TO FORMATION TYPE

Formation	Tooth	Degrees
Soft	Long, blunt chisel	2-3
Medium-soft	Long, sharp chisel	2-3
Medium shales	Medium-length chisel	1-2
Medium limes	Medium-length chisel	1-2
Medium-hard	Short chisel	0
Medium	Short projectile	0
Hard chert	Conical or hemispherical	0
Very hard	Conical or hemispherical	0

SOURCE: Jack C. Estes, "Selecting the Proper Rotary Rock Bit," *Journal of Petroleum Technology,* November 1971 (Copyright SPE-AIME 1971.)

When the drilling rate with a roller cone bit falls below a given value – say, 3–5 feet per hour – a *diamond bit* may be the best alternative. A diamond bit may cost up to four or five times as much as an insert bit – which is, in turn, a great deal more costly than a milled-tooth bit – but under certain conditions it can cost less per foot of hole than other bits. The diamonds must be kept clean and cool, so hydraulic effort is just as important for the performance of diamond bits as for roller bits. If properly used in a clean hole, a diamond bit will stay in good condition several times longer than a roller cone bit. Furthermore, up to 50 percent of the bit's cost can be recovered by salvaging the diamonds.

Diamond bits function like *drag bits* in that both weight and rotary speed are directly related to drilling speed. Recent improvements have made the drag bit a reasonable alternative to the widely used roller cone and diamond bits. New drag bits contain compacts of diamond and tungsten carbide. These polycrystalline diamond (PCD) bits (fig. 6.1) combine the diamond's hardness and abrasion resistance with tungsten carbide's durability to make long bit runs. In certain drilling situations, the PCD bit's faster penetration rate, longer time in the hole, and fewer trips result in significant cost savings over a conventional bit.

Figure 6.1. Polycrystalline diamond (PCD) bit (*Courtesy of Strata Bit*)

Bit Performance Evaluation

Bit wear evaluation. A decrease in penetration rate, reduction in drill pipe torque, or other indications of a dull bit make it necessary to trip out so that the driller or toolpusher can examine the bit (fig. 6.2). By "reading" the dull bit, an experienced toolpusher can see firsthand how successful the drilling program is. Even a bit well-suited for the formation will show some wear, and how it wears is valuable information. Improper bit selection for the formation may cause inner bearing failure, tooth or insert breakage, or wear that is faster on teeth in the inner rows than on the gauge teeth.

Improper drilling practices are evident in wear on bit cones. Skid marks on cones can indicate balling of the bit due to insufficient hole cleaning. A locked cone is a sign of outer bearing failure, usually indicating excessive weight and rotary speed. Severe cone erosion—sometimes resulting in a cracked cone or loss of inserts—is usually caused by excessive drilling fluid flow. If not fished out, lost or broken inserts can accelerate bit damage. Higher bit weight dulls bits faster (fig. 6.3) and wears out bearings faster (fig. 6.4).

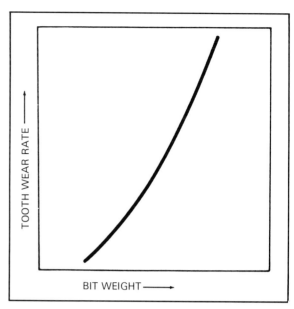

Figure 6.3. Effect of bit weight on tooth wear rate

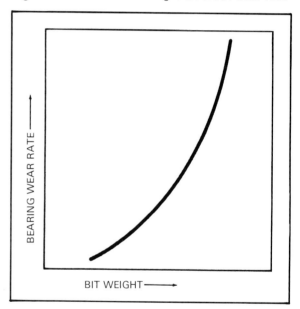

Figure 6.4. Effect of bit weight on bearing wear rate

Figure 6.2. Reading dull-bit appearance

TABLE 6.3

SAMPLE DRILL-OFF TEST
IN SANDY SHALE

WEIGHT INTERVAL (1,000 lb)	DRILL-OFF TIME (seconds)
70-65	26
65-60	26
60-55	25 (Best Run)
55-50	28
50-45	31
45-40	34
40-35	36
35-30	52
30-25	70

Drillers have long been aware that the rate of penetration is a good indicator of bit condition—that is, whether it is sharp or dull or whether the weight, rotary speed, and hydraulic effort being used are suitable for the formation. From periodic observations of the time needed to drill a foot, ten feet, or the length of the kelly, the driller determines drilling rate in feet per hour, the usual basis for comparing bit performance. Many rigs are equipped with drilling rate recorders, which enable quick review and comparison of penetration rates. If automatic devices are not available, the driller checks the time required to drill a foot or more by marking the kelly and drilling ahead with constant weight on bit. The time required to drill the selected interval is noted and compared with earlier drilling time tests. This easily understood method is the most common way of judging performance. It usually suffices for rough estimates of bit performance.

Drill-off tests. A more precise method of measuring penetration rate is the *drill-off test,* in which the driller

1. slacks off a given amount of weight on the bit;
2. ties down the brake; and
3. circulates and rotates at a constant rate.

The bit is allowed to drill off some of the applied weight. As the bit penetrates, less and less of the drill string weight is supported by the hole bottom, and more and more weight is suspended from the hook, increasing drill string tension. The increased tension causes the drill string to stretch and continue making hole, although with less and less weight on the bit. The number of seconds needed to drill off a given amount of weight, usually 2,000 to 5,000 pounds, is carefully noted, as shown by the following example (table 6.3).

Hole size was 9⅝ in., depth 10,360 ft; drill string consisted of 630 feet of drill collars and 9,730 feet of 5-in. OD drill pipe; rotary speed was 130-135 rpm and bit hydraulic power 434 hhp. The best run was 25 seconds with about 60,000 lb weight on bit.

The results of a drill-off test can be used to calculate the penetration rate. First, the driller makes a pickup and slack-off chart by marking the kelly, picking up on the drill string, and measuring how much the drill string stretches when a given amount of weight is picked up (fig. 6.5). Kelly movement per 1,000 lb picked up (or slacked off) is calculated as follows:

$$D = \frac{H_2 - H_1}{W_2 - W_1}$$

where
- D = penetration (ft per 1,000 lb slacked off);
- H_1 = height of kelly before picking up (ft);
- H_2 = height of kelly after picking up (ft);
- W_1 = suspended weight before picking up (1,000-lb increments); and
- W_2 = suspended weight after picking up (1,000-lb increments).

The weights and heights used are taken from the graph. For accuracy, end points are not used—that is, the points of maximum and zero weight on bit.

For the example shown in figure 6.5,

$$D = \frac{2.9 - 0.5}{230 - 190}$$

$$= \frac{2.4}{40}$$

$$= 0.06 \text{ ft}/1,000 \text{ lb}.$$

That is, the bit penetrates 0.06 ft for every 1,000 lb slacked off.

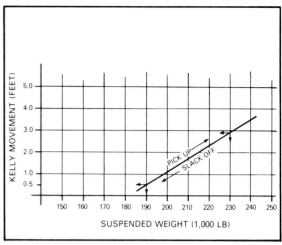

Figure 6.5. Pickup and slack-off chart

The above information can then be used to calculate the best penetration rate, using the following formula:

$$R = 3,600 \times \frac{D \times W}{T}$$

where
- R = penetration rate (ft/h);
- D = penetration (ft per 1,000 lb slacked off);
- W = weight drilled off (1,000-lb increments); and
- T = time to drill off (s).

Using the best run (25 s to drill off 5,000 lb), penetration rate can be calculated as follows:

$$R = 3,600 \times \frac{0.06 \times 5}{25}$$

$$= 43.2 \text{ ft/h}.$$

Drill-off tests can also be used to determine the optimum rotary speed by varying rpm while holding weight on bit and hydraulic effort constant. Similarly, the effect of increased hydraulic power at the bit can be measured by varying that factor while holding bit weight and rotary speed constant. To cancel out the effects of small variations in formation characteristics, several tests should be conducted and averaged.

Footage costs. Footage per bit and penetration rate have long been the accepted yardsticks for evaluating bit performance. These are two different measurements that must be combined on an economic basis to be used properly for comparison. The basic formula for determining cost per foot for a drilled interval is

$$C = \frac{B + R(T + t)}{F}$$

where
- C = drilling cost (\$/ft);
- B = bit cost (\$);
- R = rig operating cost (\$/h);
- T = drilling time (h);
- t = trip time (h); and
- F = hole drilled (ft).

Cost per foot of hole drilled, using both penetration rate and bit life, is kept low when overall rig performance is high. This type of calculation becomes very important on holes that require many bits to reach final depth. The lower cost per foot gained by changing from a usable but inappropriate bit to a more suitable bit might justify the cost of a round trip.

Suppose the driller is making 5.25 ft/h with a \$1,650 bit that is expected to last 67 hours and make 352 feet of hole. Can he expect to lower the cost per foot of hole by tripping out and changing to a \$2,811 bit that makes 9.36 ft/h and will last 81 hours and make 758 feet of hole? (Operating cost is \$110/h and trip time is 2 hours.)

1. Old bit:
$$C_1 = \frac{B_1 + R(T_1 + t)}{F_1}$$
$$= \frac{1{,}650 + 110(67 + 2)}{352}$$
$$= \$26.25/\text{ft}.$$

2. New bit:
$$C_2 = \frac{B_2 + R(T_2 + t)}{F_2}$$
$$= \frac{2{,}811 + 110(81 + 2)}{758}$$
$$= \$15.75/\text{ft}.$$

Clearly the cost per foot of hole will be reduced by changing bits.

WEIGHT ON BIT AND ROTARY SPEED

Factors Related to Bit Type

Achieving optimal drilling rates requires coordinating the mechanical factors of bit weight and rotary speed with bit selection. As long as bit hydraulics provides proper bottomhole cleaning, increasing either the bit weight or the rotary speed usually increases the rate of penetration. However, weight on bit and rotary speed must be increased carefully to avoid undue bit wear or unwanted hole deviation. The added cost of correcting a crooked hole or replacing a prematurely worn bit can quickly offset the intended benefits of high-energy drilling. Optimization usually requires bit weight and rotary speed to be changed inversely—that is, when one is increased, the other is decreased. Table 6.4 demonstrates this inverse relationship and shows suggested weight and speed combinations for a medium- to soft-formation insert bit.

TABLE 6.4
RECOMMENDED WEIGHT AND SPEED FOR TYPICAL INSERT BIT IN MEDIUM-SOFT FORMATIONS

FORMATION	BIT DIAMETER (inches)	HIGH SPEED/LOW WEIGHT		LOW SPEED/HIGH WEIGHT	
		Rotary Speed (rpm)	Weight on Bit (1,000 lb)	Rotary Speed (rpm)	Weight on Bit (1,000 lb)
Shale	6¼ - 6½	70	15-16	55	20-21
	7⅞	70	23	55	32
	8⅜ - 8¾	70	27-28	55	34-36
	9½ - 9⅞	70	30-32	55	39-40
	11	70	33	55	44
	12¼	70	34	55	44
Limestone, dolomite	6¼ - 6½	65	20-21	45	27-28
	7⅞	65	31	45	39
	8⅜ - 8¾	65	34-35	45	42-44
	9½ - 9⅞	65	38-40	45	48-49
	11	65	43	45	50
	12¼	65	47	45	53
Unconsolidated shales, limestones, and sands	6¼ - 6½	55	20-21	40	25-26
	7⅞	55	31	40	35
	8⅜ - 8¾	55	34-35	40	39-40
	9½ - 9⅞	55	38-40	40	44-45
	11	55	43	40	50
	12¼	55	47	40	54

SOURCE: J. H. Allen, "Determining Parameters That Affect Rate of Penetration," *Oil and Gas Journal*, October 3, 1977, Table 2 (Copyright 1977 Pennwell Publishing Company).

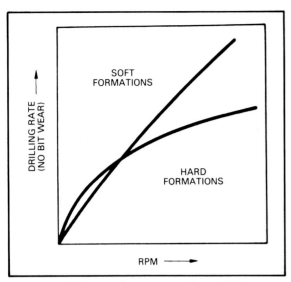

Figure 6.6. Effects of rotary speed on drilling rates

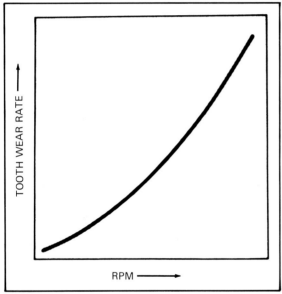

Figure 6.7. Effect of rotary speed on tooth wear

Of course, different types of formations respond differently to increases in bit weight and rotary rpm. For example, penetration rates in most soft, shaly formations increase with rotary speed (fig. 6.6). The shearing action of soft-formation bits can be tailored for rotary speeds up to 250 or 300 rpm. However, increasing rotary speed also increases bit tooth wear (fig. 6.7). Increases in bit weight may increase the rate of penetration; however, the chances for hole deviation also increase, especially as the bit becomes dull.

Improvements in bit tooth and bearing construction have increased the permissible weight on bit in hard formations. Heavy bit weights (5,000–10,000 lb/in. bit diameter) are required to overcome the compressive strength of the rock. However, high rotary speeds can result in shock loads that will shatter bit teeth and cause premature failure in the threaded connections of the drill string. Increasing rotary speed has less effect on penetration rate in hard formations than in soft formations (fig. 6.6).

Although some drillers consider weight on bit the most important factor in making hole, excessive weight may result in hole deviation unless the bit and drill collars are properly stabilized. In tilted formations or those with alternating soft and hard streaks, hole deviation can be a problem even with light bit loads. The best way to drill a vertical hole is with good bottom-cleaning hydraulics and a sharp bit, giving adequate penetration even with light bit weight. Proper use of stabilizers and large-diameter drill collars can also reduce the likelihood of doglegs or unwanted deviation in the wellbore. Proper stabilization also helps bits run true, improving bearing life and reducing tooth wear.

Rig Capabilities

Rig capabilities must be considered before increasing bit weight and rotary rpm. Additional weight means replacing drill pipe with drill collars, which not only cost more to buy and maintain but also require extra trip time for handling. Safety clamps and lifting subs must be used, requiring more work at the rotary. Even though each stand is handled in 5 minutes, 2 hours may be required to break out and make up an assembly of thirty or more collars during a round trip.

Hoisting time for round trips and connections (essential for changing bits, adding pipe, and changing components of the drill collar string) takes time away from the rig's main purpose. Increasing drill collar weight increases the hoisting requirements of the rig. Figure 6.8 shows the hook power needed for pulling various loads at different speeds. This chart shows the power required at the hook. Additional engine power is needed to overcome mechanical losses due to friction in the compound, drawworks, wire ropes, and blocks. The usual arrangement using several engines loses 15% or more of its power between the engines and the hoisting drum. Another 15% is lost between the drum and the hook. The chart shows that about 700 hook horsepower is needed to pull a load of 300,000 pounds at 75 feet per minute. Engine power required for this load with ten lines strung is approximately 1,000 hp. Power for hoisting becomes more significant on large or deep holes requiring heavy drill collars and drill pipe and on holes where the formation dulls bits quickly, requiring many trips.

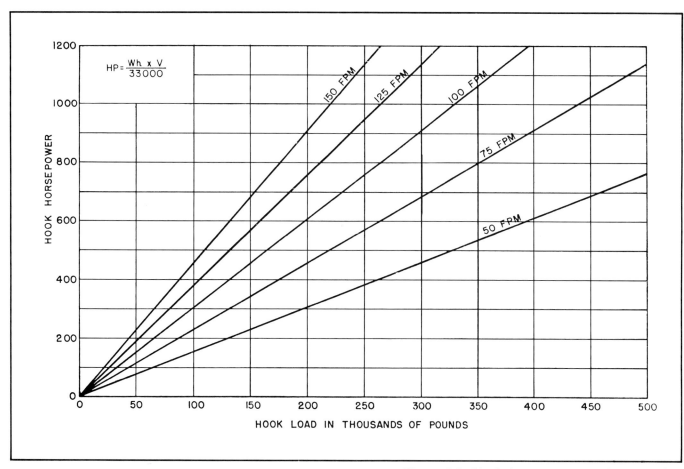

Figure 6.8. Hook horsepower requirements for hoisting (*API Bulletin D-10*)

Rotary torque, or resistance to turning, also increases as weight on bit and rpm are increased, requiring more horsepower to operate the rotary (fig. 6.9). The combined factors of high rotary speed, rotary torque, and weight on bit can impose tremendous stress on the drill string. Although drill pipe failure is more likely to result from metal fatigue (due to notches, imperfections, or corrosion) than from torsional loads, the driller must be aware of the stress developed by rotary torque and how it may affect the equipment. At high speed, a lot of energy is contained in the rotating mass of the drill string. A sudden seizure, or lockup, of the bit can severely damage the drill string through tool joint failure or twistoff. Furthermore, if the drill string does not run true at high speed, increased hole friction and vibration may cause bit bearing failure and tooth breakage.

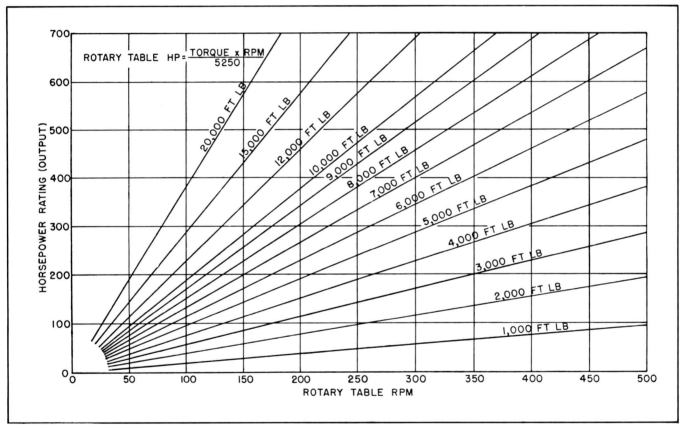

Figure 6.9. Rotary table horsepower requirements
(*API Bulletin D-10*)

The type of bit used will also affect the amount of rotary power needed. Long-tooth bits require more horsepower than do short-tooth bits for the same formations. Generally, steel-tooth bits can withstand more power (weight and rotary speed) without chipping or breaking than can insert bits (table 6.5); however, improved insert grades and shapes allow higher rotary speeds in soft formations and higher weights in hard formations than were previously possible. Diamond bits operate best at low bit weights and high rotary speeds; the ideal load for a diamond bit is the lowest weight at which good penetration can be achieved. Figure 6.10 shows the recommended maximum weights for diamond bits. Rotary speeds are usually around 100 rpm; however, much higher speeds can be used with downhole motors.

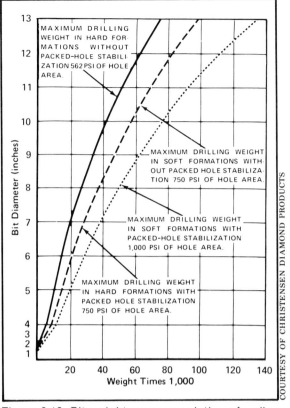

Figure 6.10. Bit weight recommendations for diamond bits

TABLE 6.5

Weight and Speed Ranges
For Roller Cone Bits

Formation Type	Weight on Bit (psi bit diameter)		Rotary rpm	
	Milled-tooth Bits	Insert Bits	Milled-tooth Bits	Insert Bits
Soft	3,000-7,000	3,000-5,000	60-250	35-100
Medium	4,000-9,000	4,000-6,000	40-100	35-60
Hard	5,000-10,000	4,500-7,000	35-80	35-50

DRILLING FLUIDS

Mud

As many drillers would agree, the drilling fluid that makes hole fastest is clear water. Unfortunately, clear water is not always the best drilling fluid. Drilling mud performs other functions vital to cost-effective drilling. The idea is to establish and maintain a drilling fluid program that effectively deals with variations in formation properties and mechanical factors affecting the drilling rate.

Mud programs not only vary from one well to another, but also change during the drilling of any single well. Gas-charged sands, sloughing shales, unexpected high-pressure zones, high temperatures, and other factors often require a drilling fluid with special properties. The mud properties with the greatest effect on penetration rate are density, solids content, viscosity, water loss, and oil content.

High-density (heavy) mud is a natural enemy of fast drilling. Low-density drilling fluid permits faster drilling because it imposes less hydrostatic pressure on bottom. Low hydrostatic pressure allows chips to be formed with less weight on the bit and lower rotary power. When mud density is excessive, high differential pressure between the mud column and the formation causes a *chip hold-down effect* (fig. 6.11), and the bit drills the same material over and over. In such cases, mechanical energy on the bit must be increased to avoid a substantial drop in the drilling rate.

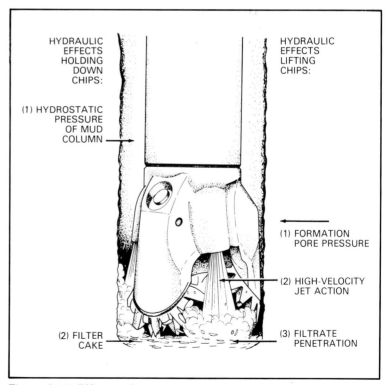

Figure 6.11. Effects of hydraulics on hole bottom

Field experience has shown that the fastest drilling progress is accomplished with the lightest possible drilling fluid, other factors being equal. Figure 6.12 illustrates the effect on drilling rates of reducing mud weight from 10.4 to 9.6 ppg. Rotating time to 9,000 feet was lowered from about 550 to 250 hours. Besides reducing drilling time, light mud costs less to buy and maintain than heavy fluid.

Light mud not only permits faster drilling than heavier fluid, it may also reduce the likelihood of lost circulation, wall-stuck pipe, swabbing, and pressure surges. However, it also increases the risk of a blowout from an abnormally pressured formation, one of the most dangerous and expensive things that can go wrong with a well. This risk, plus the risk of losing valuable drilling time fighting a kick, must be balanced against the economic savings of drilling underbalanced—that is, with mud hydrostatic pressure lower than formation pressure.

Solids content. Low-solids fluid—water, where feasible—promotes faster penetration than lightweight mud under the same conditions of bit weight, rotary speed, and hydraulics (fig. 6.13). Low-solids mud is used principally with mud weights below 10 ppg and circulation rates high enough to lift cuttings out of the hole. Where conditions permit, holes are sometimes drilled to several thousand feet with just clear water.

Weighting material in the mud can slow the rate of penetration by contributing to the hold-down effect. Small particles of the material can plug the fractures where a chip has been sheared from the hole bottom, delaying fluid pressure equalization beneath the chip and inhibiting its removal. Low-solids mud forms less *wall cake* (buildup of solids on the hole wall) to hold down chips, and loses more fluid to formation, helping to separate the chips from the hole bottom.

Low-solids fluid is usually maintained by circulating it through the reserve pit to allow fine solids to settle out, or by using centrifuges, desilters, desanders, and mud cleaners. Mud chemicals that cause the particles to coagulate and settle out can also be used.

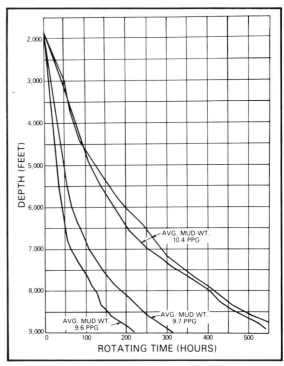

Figure 6.12. Effect of mud weight on drilling rate

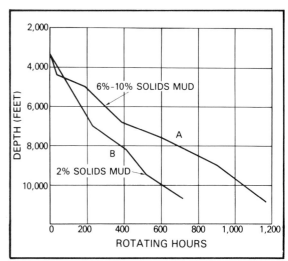

Figure 6.13. Effect of mud solids content on drilling rate

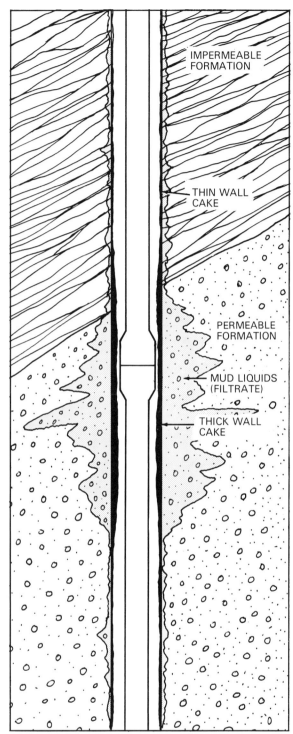

Figure 6.14. Fluid loss in permeable formations

Viscosity. Increases in *viscosity,* the mud's resistance to flow, increase circulation pressure losses, reducing bit hydraulic horsepower (BHHP) and making bottomhole cleaning more difficult. Highly viscous or thick muds tend to hold chips on bottom more than low-viscosity muds, reducing the penetration rate. *Gel strength,* a property that slows the settling of cuttings when circulation is stopped, should also be maintained to keep the hole clean.

Fluid loss. The pressure of the mud column may force some of the liquid component of the mud into the more permeable formations (fig. 6.14). This water loss, or *filtration loss,* is beneficial in two ways. The initial loss, called *spurt loss,* can make formations easier to drill by helping to separate chips from the formation. Filtration also causes a buildup of wall cake (sometimes called *filter cake*), sealing the wellbore and preventing the leakage of whole fluid into the formation. However, excessive fluid loss can cause problems: wall cake buildup makes the drill string more likely to become wall-stuck, and oil recovery may be hampered by filtrate in producing formations.

Oil content. Adding oil to water-based muds can increase penetration rates in formations where high temperatures, sloughing shales, or pipe sticking are expected. By minimizing hole friction, the oil reduces the likelihood of pipe sticking, helps keep the bottomhole assembly rotating freely, and effectively increases bit weight. Oil also keeps the bit from balling up in certain hydratable clays and shales.

Air or Gas

In areas where formations do not contain too much water and will not slough, air or gas drilling permits the fastest penetration rates. The lack of drilling fluid hydrostatic pressure eliminates or even reverses the holddown effect, causing cuttings literally to explode from the bottom of the hole after being barely touched by the bit teeth, even with a dull bit. Footage per bit and overall penetration rate are usually much better with air or gas than with water or mud. Problems with lost circulation are avoided and casing can be set shallower.

Unfortunately, there are relatively few places where air or gas drilling are feasible. A small amount of formation water may turn the returning stream of air and cuttings to a thick, adhesive mud. Additional water or foaming agents may have to be used, increasing drilling costs. Special equipment needed for air or gas drilling may cancel rig time savings. As in all drilling operations, the total cost must be considered.

BIT HYDRAULICS

Hydraulic Horsepower

The hydraulic horsepower of the circulating fluid at the bit is vital to rotary bit performance. Increasing bit weight and rotary power does not proportionally increase penetration rate unless hydraulic horsepower is sufficient to clean the hole bottom and move cuttings to the surface. Penetration rates are optimized by using, within rig capabilities, appropriate combinations of weight, rotary speed, and hydraulic power.

Field results show that additional hydraulic power at the bit can be converted into faster drilling if additional weight is put on the bit. Figure 6.15, for example, shows the results of a test using a 9⅞-inch bit in a sandy shale at 7,500 feet. For each hydraulic horsepower input to the pump, weight on bit was varied; the rotary was run at whatever speed was required to get the maximum rate of penetration. Table 6.6 shows the resulting relationships among bit hydraulic power, weight, and penetration.

With 200 hydraulic horsepower available at the bit, the best penetration rate (23 ft/h) was produced with 34,000 lb on the bit; and with 400 hhp, the optimum weight (54,000 lb) produced 52 ft/h.

The formula for bit hydraulic horsepower—

$$H = \frac{P \times Q}{1,714}$$

where

- H = hydraulic horsepower;
- P = pressure (psi); and
- Q = flow (gpm)—

shows that hydraulic horsepower depends upon both pressure and flow rate. Friction inside the drill stem does not reduce the flow rate because the same amount of fluid must leave the drill stem as is pumped in. However, friction does reduce the pressure as the fluid travels down the drill string. Therefore, the percent of hydraulic horsepower lost between pump and bit is the same as the percent loss of pressure. In a typical well, about a third of the hydraulic power is lost between the pump and the nozzles.

Pressure loss depends upon many factors: the length and inside diameter of the drill string, friction between the drill pipe and the drilling fluid, fluid density and viscosity, and

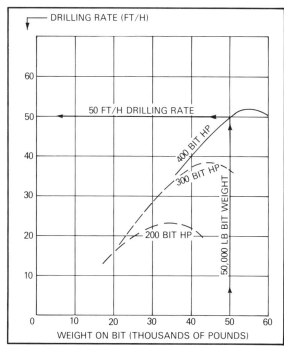

Figure 6.15. Effect of bit hydraulic power on drilling rate

TABLE 6.6
RELATIONSHIPS OF BIT HYDRAULIC POWER
TO WEIGHT ON BIT AND PENETRATION RATE

Bit Hydraulic Power	Optimum Weight on Bit	Optimum Penetration Rate
200 hhp	34,000 lb	23 ft/h
300 hhp	44,000 lb	39 ft/h
400 hhp	54,000 lb	52 ft/h

flow rate. The higher the flow rate, the greater is the pressure lost to friction in the drill string. Therefore, for a given pump output power, a high-volume, low-pressure discharge will lose more pressure, and therefore more power, in the drill string than will a low-volume, high-pressure discharge.

The allowable pump discharge pressure depends on the pump liner size and the corresponding pressure rating. Using larger liners and slowing the pump—but maintaining the same pressure at the pump discharge—reduces system pressure loss. Operating pumps this way results in a lower volume of flow but less pressure loss between pump and bit and therefore greater efficiency. Of course, pressure must not exceed the pump rating, nor can annular return velocity be allowed to fall below the minimum required for effective hole cleaning; otherwise, the benefits of more efficient delivery of power to the bit are lost.

Other conditions limiting the hydraulic power that can be pumped to the bit are well depth, drill pipe size, and mud density. For a given pump output power, less power reaches the bit as the well is drilled deeper because each additional joint of pipe adds to the frictional loss in the drill string. Small-diameter drill pipe causes more turbulence, and therefore more friction loss, than large-diameter pipe. Using plastic-lined pipe, lower-density mud, and mud containing oil help reduce fluid friction.

Nozzle Selection

Hydraulic power is put to work by the bit nozzles, which hold back-pressure on the fluid. By constricting the path of mud flow, the bit nozzles concentrate the circulating fluid into powerful streams that are vital to bottomhole cleaning.

Proper nozzle selection is essential to a good hydraulics program. By determining the initial pressure of the mud pumps, calculating the pressure losses throughout the system, and referring to tables, the most efficient combination of nozzle sizes can be selected. With a constant pump flow rate, an increase in nozzle size will decrease the mud velocity at the bit. For example, if too much pressure is obtained with a given pump speed and liner size arrangement, one of the nozzles in the bit can be changed to the next larger size; if too little pressure, a nozzle should be pulled and a smaller size installed.

A suitable nozzle combination uses most of the hydraulic power available at the bit, leaving just enough to circulate the cuttings up the annulus. In general, of course, the optimum combination is the one that results in the highest drilling rate.

FORMATION PROPERTIES

The nature of the formations to be drilled has a great deal to do with the drilling rate. The geological factor is a drilling condition that the driller cannot change. However, his job can be much easier and safer if he knows as much about the formation as possible. Offset wells can provide valuable information on the characteristics of the formation being drilled; petroleum geologists can help the driller plan the drilling program.

The compressive strength of a rock is its ability to resist physical stress. In hard formations, weight on bit must be sufficient to overcome compressive strength so that the rock will fail under the applied stress of the drilling mechanism. Compressive strength generally increases with depth.

Higher porosity usually means lower compressive strength; porous formation zones drill faster than nonporous zones because a porous rock structure fails more readily when drilled. Hydrostatic and formation pressures tend to equalize in permeable formations. This tendency minimizes or eliminates the chip hold-down effect and contributes to faster drilling.

The reaction of formation material as it comes into contact with drilling mud can reduce the drilling rate. For example, when water-base muds are used, some clays and shales tend to form a sticky mixture that can make bottomhole cleaning difficult and ball up the bit.

In addition to drillability, other geological factors should be considered before drilling begins. Formation fluids and pressures are a critical concern. The driller should know whether and at what depths to expect abnormal pressures and plan his mud program and blowout prevention needs accordingly. Saltwater flows may require special maintenance and conditioning of the drilling mud.

Drilling in crooked-hole country involves another set of concerns. Besides knowing the location and severity of crooked-hole zones, the driller must also be aware of the operator's policy, contract limitations, and legal restrictions on wellbore deviation. Drilling operations must conform to contract specifications on both total hole angle change and maximum deviation in any section.

Lost circulation zones can be found at any depth in almost any geological area. Not only do they disrupt drilling while circulation is being restored, they often cause differential pipe sticking, requiring a fishing job. Although it is impossible to drill around lost circulation zones, preventive measures can be taken before drilling into them.

Sloughing or heaving shale is another type of unstable formation that can slow drilling progress. When wetted by water in the drilling mud, these formations swell and fill the wellbore. They usually require drilling with less weight, faster rotation, and higher circulation rates. If not drilled carefully, sloughing shale can stick the drill string and result in a fishing job.

COMPUTERIZED OPTIMIZATION

More drilling operations are being optimized by computer than ever before. Computer programs for achieving the maximum cost-effective penetration rate are used in place of programmable calculators, slide rules, tables, and graphs to specify the most suitable bit type, drilling weight, rotary speed, mud program, and hydraulics for a given drilling situation. Computers can process much more data, and much faster, than manual optimization techniques.

A drilling plan for a proposed well or for a well in progress can be formulated by using data from an offset well (a nearby well that was drilled under similar conditions and through the same formations). A typical computer analysis consists of a series of programs.

1. The first program analyzes bit records, mud reports, and formation logs from the offset well and determines whether another type of bit would be more cost-effective in the proposed well.

2. A hydraulic analysis program determines flow rate and other pertinent data.

3. Another program optimizes the hydraulics for the proposed mud weight. This program specifies jet nozzle sizes, pump volume, and circulating pressure.

4. Hydraulics in the proposed well is compared with hydraulics in the offset well to determine further necessary adjustments in mud weight and composition, pump volume, and nozzle sizes.

5. A program is run to determine if penetration rate and cost effectiveness can be improved by changing weight on bit and rotary speed.

6. Finally, with all other drilling factors optimized, a bit selection program further refines bit selection to take full advantage of the drilling program.

When drilling is underway, various programs can be used to adjust drilling parameters as conditions change. Information from many wells can be transmitted electronically to a central location for analysis, and results can be returned almost instantaneously to individual drilling operations. With an automatic telecommunications link, for instance, an offshore driller has access to both raw data on the rig and detailed analyses from a powerful shore-based computer.

Drilling conditions change continuously during the progress of a well. Mud weight has to be adjusted in drilling high-pressure formations; drill string length, weight, and diameter vary with well depth; available pump pressure falls off as the hole approaches maximum depth. Optimized drilling means keeping the operation at its most efficient from the beginning to the end of a hole. Because of its speed, accuracy, and increasing cost effectiveness, the use of a computer will almost inevitably become standard practice in drilling as it has in other technologies.

LESSON 6 QUESTIONS

Put the letter for the best answer in the blank before each question.

Questions 1–17 cover material found in the sections entitled **Introduction** and **Bits.**

_____ 1. The goal of optimization is—
   A. to make the bit last longer by using the slowest possible rotation.
   B. to drill the most usable hole at the lowest cost.
   C. to drill the straightest and largest hole possible.
   D. to use the least expensive equipment available.

_____ 2. Factors affecting rate of penetration include all of the following *except*—
   A. weight on bit.
   B. magnetic declination.
   C. rotary speed.
   D. type of bit selected.

_____ 3. Of the factors involved in making hole, the one that the driller cannot control is—
   A. bit hydraulics.
   B. formation properties.
   C. drilling fluid properties.
   D. rotary speed.

_____ 4. According to tables 6.1 and 6.2, the best bit to use in medium-hard formation is—
   A. a hardfaced-tip milled tooth bit with 3° cone offset.
   B. a short-chisel insert bit with 0° cone offset.
   C. a conical-tooth insert bit with 4° cone offset.
   D. a long, sharp-chisel insert bit with 2° cone offset.

_____ 5. Soft-formation rock bits have longer teeth than hard-formation bits. (T/F)

_____ 6. A roller cone bit with cones offset 4° is designed for—
   A. soft formations.
   B. medium formations.
   C. hard formations.
   D. any type of formation.

_____ 7. Hard-formation drag bits drill only with slow rotation. (T/F)

_____ 8. Which of the following types of bits makes hole principally by crushing the formation?
   A. A drag bit
   B. A diamond bit
   C. A milled-tooth roller cone bit with 3° cone offset
   D. A roller cone insert bit with 0° cone offset

_____ 9. What is the principal advantage of insert bits over milled-tooth bits?
   A. Insert bits are less expensive.
   B. Insert bits can withstand faster circulation rates.
   C. Insert bits last longer.
   D. Insert bits can be salvaged for 80% of their initial cost.

_____ 10. Although initially more expensive, a diamond bit may make hole more economically than other bits because—
   A. it lasts longer.
   B. it requires fewer persons on the rig.
   C. it can take unlimited weight on bit.
   D. it is used without drilling mud.

_____ 11. Drag bits are no longer used in the oil patch. (T/F)

_____ 12. A locked bit cone may indicate—
   A. excessive weight on bit.
   B. insufficient rotary speed.
   C. excessive circulation rate.
   D. inadequate torque.

_____ 13. The driller can determine penetration rate by—
   A. drilling ahead with the pumps shut down.
   B. checking rotary speed with a stopwatch.
   C. conducting a drill-off test.
   D. studying bit wear patterns.

_____ 14. A driller conducts a pickup and slack-off test. With the bit on bottom and a 140,000-lb hook load, he marks the kelly. Picking up on the drill string, but with the bit still on bottom, he increases hook load to 195,000 lb. The mark on the kelly rises 1.8 ft. How much does the drill string stretch? (Use the formula on p. 198.)
   A. 0.01 ft/1,000 lb
   B. 0.03 ft/1,000 lb
   C. 0.08 ft/1,000 lb
   D. 0.18 ft/1,000 lb

_____ 15. Calculate the penetration rate (drilling rate), given the following information: penetration 0.05 ft/1,000 lb slacked off, 2,000 lb drilled off in 12 s. (Use the formula on p. 199.)
   A. 1.5 ft/h
   B. 12 ft/h
   C. 15 ft/h
   D. 30 ft/h

_____ 16. What is the cost per foot of hole drilled using a $1,500 bit that lasts 55 h and drills 470 ft of hole on a rig costing $90/h to run and requiring 1.5 h for a round trip? (Use the formula on p. 200.)
   A. $3.19/ft
   B. $10.53/ft
   C. $14.01/ft
   D. $186.07/ft

_____ 17. A driller must choose between two bits for a particular formation. Bit A costs $875 and is expected to last 45 h and drill 387 ft. Bit B costs $2,340 and should last 58 h and drill 563 ft. The rig costs $125/h to operate; trip time at this depth is 2.5 h. Which bit is more economical to use? (Use the formula on p. 200.)
A. Bit A.
B. Bit B.
C. They cost about the same per foot of hole.

Review the section entitled **Weight on Bit and Rotary Speed.** Then answer questions 18–24.

_____ 18. If weight on bit is increased, rotary speed should usually be –
A. increased.
B. decreased.
C. kept the same.

_____ 19. According to table 6.4, about how much weight should be applied to a 9½-inch insert bit turning at 45 rpm in dolomite?
A. 48,000 lb
B. 44,000 lb
C. 40,000 lb
D. 35,000 lb

_____ 20. Attempting to increase the rate of penetration by simply increasing weight on bit and rotary speed can result in –
A. higher circulation rates at the bit.
B. hole deviation.
C. excessive penetration rates.
D. all of the above.

_____ 21. Increasing rotary speed has less effect on penetration rates in hard formations than in soft formations. (T/F)

_____ 22. Look again at figure 6.8. How much hook horsepower is needed to lift 250,000 lb at 50 ft/min?
A. 390
B. 530
C. 760
D. 1,140

_____ 23. According to table 6.5, would it be wise to apply 10,000 lb/in. of bit diameter to an insert bit turning at 50 rpm in hard formation?
A. Yes
B. No

_____ 24. In any given formation, a milled-tooth bit is usually rotated faster than an insert bit. (T/F)

To answer questions 25–34, review **Drilling Fluids** and **Bit Hydraulics**

_____ 25. Of the following drilling fluids, the one that makes hole fastest is –
   A. oil-based mud.
   B. high-density mud.
   C. clear water.
   D. bentonite.

_____ 26. Reducing mud density is likely to –
   A. decrease penetration rate.
   B. reduce the risk of a blowout.
   C. increase penetration rate.
   D. increase mud maintenance costs.

_____ 27. In the example shown in figure 6.12, drilling to 7,000 ft with 9.6-ppg mud took _____ than with 10.4-ppg mud.
   A. 50 h longer
   B. 50 h less
   C. 180 h less
   D. 320 h less

_____ 28. Using low-solids drilling fluid –
   A. reduces wall-cake buildup.
   B. increases the chip hold-down effect.
   C. reduces circulation rate requirements.
   D. increases bit wear.

_____ 29. Differential pressure from wellbore to formation is least using _____ as the drilling fluid.
   A. high-viscosity mud
   B. oil-base mud
   C. clear water
   D. air

_____ 30. In the example shown in figure 6.15, which of the following combinations of weight on bit and bit hydraulic horsepower produced the highest penetration rate?
   A. 40,000 lb, 200 bhhp
   B. 30,000 lb, 200 bhhp
   C. 30,000 lb, 300 bhhp
   D. 23,000 lb, 300 bhhp

_____ 31. If circulating pressure at the bit is 1,400 psi and flow rate is 430 gpm, about how much hydraulic horsepower is available at the bit? (Use the formula on p. 209.)
   A. 430
   B. 600
   C. 1,714
   D. 350

_____ 32. The amount of hydraulic power that can be pumped to the bit increases as _____ increase.
A. well depth and temperature
B. mud density and rotary speed
C. drill pipe ID and mud oil content
D. casing ID and weight on bit

_____ 33. With a constant pump flow rate, increasing nozzle size _____ mud velocity at the bit.
A. decreases
B. increases
C. does not affect

_____ 34. The best nozzle combination uses all of the hydraulic power available at the bit. (T/F)

Finally, answer questions 35–40 based on your reading of **Formation Properties** and **Computerized Optimization.**

_____ 35. The compressive strength of rock generally increases with–
A. porosity.
B. depth.
C. weight on bit.
D. hydrostatic pressure.

_____ 36. All else being equal, the easiest formation to drill is one that is–
A. nonporous and nonpermeable.
B. nonporous and permeable.
C. porous and nonpermeable.
D. porous and permeable.

_____ 37. The safest thing to do with lost circulation zones is to drill around them. (T/F)

_____ 38. Sloughing or heaving shale generally requires drilling with–
A. less weight on bit.
B. higher circulation rates.
C. faster rotation.
D. all of the above.

_____ 39. Computers are useful for drilling optimization because they automatically collect all required input information. (T/F)

_____ 40. Using the computerized optimization technique described in the text, which of the following data would the driller have to obtain first?
A. Hydraulics in the proposed well
B. The size of the petroleum reservoir
C. Offset well bit records
D. Permeability of the producing formation